Guide to where the wildflowers hide

The Joy of Looking

Bruce Partridge

author of
Borderline Hardy in 5b

The Joy of Looking: Guide to where the wildflowers hide
© 2023 Bruce Partridge

Cover image: Tom Young for the Seawall Trail Society
Cover design: Rebekah Wetmore
Editor: Andrew Wetmore

ISBN: 978-1-998149-19-3
First edition November, 2023

MOOSE HOUSE
PUBLICATIONS

2475 Perotte Road
Annapolis County, NS
B0S 1A0

moosehousepress.com
info@moosehousepress.com

We live and work in Mi'kma'ki, the ancestral and unceded territory of the Mi'kmaw people. This territory is covered by the "Treaties of Peace and Friendship" which Mi'kmaw and Wolastoqiyik (Maliseet) people first signed with the British Crown in 1725. The treaties did not deal with surrender of lands and resources but in fact recognized Mi'kmaq and Wolastoqiyik (Maliseet) title and established the rules for what was to be an ongoing relationship between nations. We are all Treaty people.

Also by Bruce Partridge

Borderline Hardy in 5b, also from Moose House

Dedicated
to all who treasure our native plant heritage,
including my children, Catherine, Margaret, Savannah and Austin,
and most of all
my wife and soul-mate Mary, who tramps alongside me through
wild places, undaunted by barrens, bogs, or blackflies.

The Joy of Looking

Figure 1: Wild Nova Scotia

Introduction

The wildflowers of Nova Scotia—what do we have? What *did* we have? And where are they now?

We were taught in school that this country was discovered and claimed by European adventurers and their monarchs. They claimed the fish, then the timber; they claimed the seal, the walrus and the whale. In this frenzy of claiming, kings competed desperately, as well, to claim exotic plants for the royal gardens, sending botanists to unspoiled lands on voyages of discovery.

The first explorers to visit Nova Scotia found a virgin land of forest and flower—species unknown in Europe—apparently inexhaustible and ripe for plunder. That there were people on the land already was a discovery they chose to ignore.

The native people, who had lived on this continent for thousands of years, knew every plant and tree, and didn't have to discover anything. That they had prior claim to the land was obvious, but they were no match for the craft and wilful stupidity of the Europeans.

One must, however, admire the botanist. Usually a young man recruited from the gardening staff of an estate or botanical garden and sent to a distant land in the perilous days of sail, he might, and often did, die young in one of a variety of ways. He accepted the native people, learning their language, and displayed the spirit of adventure, love of plants, and disregard for wealth which I would like to say is characteristic of the botanist even today.

Because of its abundant timber and fish, and its relative proximity to Europe, Nova Scotia was visited and botanized early. First records were spotty and concerned mostly with the trees but exploration really got underway in the 1800s.

By the 1860s and 70s, the population was booming, Dalhousie University, Pictou Academy and other places of higher learning were established, and numerous amateur botanists combed the more accessible parts of the province, recording their finds in various scientific journals of the time.

In 1876, A.W.H. Lindsay, a professor at the Dalhousie School of Medicine and a dedicated botanist, compiled the first *Catalogue of*

Figure 2: Hidden treasures

the *Flora of Nova Scotia*. Following this was a flurry of further exploration by teachers, clergymen and scientists of various disciplines compiling lists of species found in their particular neighbourhoods.

In the 1920s, parties of professionals, chief among them M.L. Fernald of the Gray Herbarium at Harvard University, delved deeply into the swamps, bogs and seashores of Southwestern Nova

Figure 3: The joy of looking

Scotia, and the dripping cliffs of Cape Breton, demonstrating that there was still a great deal to be discovered. No one attempted to organize all this information systematically until in 1946, Albert Roland of the Nova Scotia Agricultural College, a botanist himself, authored the first edition of *The Flora of Nova Scotia* that we know and love. In 1969, a new and expanded edition of the Flora, co-written by Professor Roland and Professor E.C. Smith of Acadia University, appeared.

The Flora has continued to evolve. New plant discoveries have never ceased in the province, and many categories have been re-organized. Marian Zinck and the staff of the Nova Scotia Museum of Natural History authored a redesigned and expanded two-volume version in 1996, and since then it has gone digital.

It is rare, in 2023, to discover a new species, but botanists are fully employed re-evaluating old ones. In the field, intrepid re-searchers on foot or by canoe probe into the most out-of-the-way and inaccessible parts of the province to check up on the status of endangered species and populations. Amateur botanists and plant hobbyists, among whom my wife Mary and I include ourselves, en-joy roaming the outdoors and visiting forgotten places in our own search for surprises.

Today, the grand prize is not often the discovery of a new spe-cies, but can be the discovery of a new and hidden location for an old one. These discoveries are often made surprisingly close to home. Nova Scotia was colonized so early, relative to the rest of Canada, that there are few large areas of wilderness left, but there are plenty of little ones.

Plant exploration is never dull, if you like growing things, and takes you out into the full realm of nature, including birds and wildlife. Searching intently for anything in the natural world lifts you miraculously out of yourself, above your cares, and, as they say, into the moment. Carl Munden, in his *Native Orchids of Nova Scotia*, says it best: "Perhaps the greatest joy of all is to get out into the forest and bog on a field trip. The hunt itself is the best experi-ence and enjoyment you can have."

That is what this book is about. The Joy of Looking.

—BP, July, 2023

1: The Flora

Figure 4: Monarch butterflies on common milkweed

Since it is winter right now, and months before we can set out, let's talk about what we will be looking for and all those who were looking before us.

We know the native plants that grow in Nova Scotia, and those that grew here in the past, because botanists are compulsive record keepers, dutifully reporting their finds in letters and journals. There exists, therefore, a written trail of discoveries beginning in the 1700s, and not finished yet. What we know about our flora today is the result of over 200 years of exploration in every corner of the province, sometimes intense, other times sporadic, the work of both amateurs and professionals.

Written reports of discoveries in Nova Scotia were recorded in "Floras" such as that of A.W.H. Lindsay in 1876, and later ones by Roland and Smith and Marian Zinck. Unidentified specimens were dried and sent to botanists to be examined and classified—something that would have been impossible without Linnaeus and the key.

This key, known as a dichotomous key, is like a game where two possibilities are posed, one is selected and leads to two more, and on and on until the end of the line where the exact identity of the specimen is determined. The dichotomous key is the cornerstone of systematic botany and relies upon the system of plant classification still useful worldwide and invented by the great Carl Linnaeus of Sweden—the father of modern taxonomy—in 1735.

Figure 5: Carl Linnaeus

Before Linnaeus, the inventory of known plants of the world was vast and chaotic. There was no universal language with which to talk plants. There was no known system with which to organize them and discover their relationships. The same plant might be, and probably was, known by a different name in a different language in every country. Explorers could only describe new discoveries by using names given them by the local inhabitants and illustrations painted by artists.

Linnaeus took care of all this.

First of all, he devised principles to identify and classify a given plant by examination of its floral parts, such as the number of anthers or the placement of the ovary or the arrangement of the sepals and petals. These things are more constant and characteristic than the shapes of leaves, for instance.

Furthermore, plants with the same arrangement of flowers could be considered to be related developmentally and grouped into families and genus and species.

To overcome the problem of different languages, Linnaeus chose to name his taxonomic categories, and the plants within them, using Latin—a language equally incomprehensible to everyone. Using Latin, and the Linnaean system of classification by flower parts, botanists in every part of the world could now discuss new discoveries and know that they were talking about the same species.

Letters began to fly—or maybe I should say float, for in 1735, in the age of sail, a letter from a botanist abroad to a colleague in, say, London might take six months or a year to arrive.

Linnaeus' system of plant classification was accepted very rapidly, considering. Young botanists, students of Linnaeus at Uppsala University, fanned out to introduce his system throughout Europe and European possessions worldwide.

In Britain, Joseph Banks, a young aristocrat and enthusiastic botanist, was one of the first to sign on. In 1766 he sailed, as a botanist, with an expedition that explored Labrador and Newfoundland, and probably a bit of Nova Scotia as well. His take was an impressive haul of dried plant specimens which would be sorted and classified when he returned to Britain nine months after sailing.

Soon after, Banks was with Captain James Cook on a perilous, three-year journey around Cape Horn to Tahiti and back by way of what we now call Australia, and along the coast of Africa.

Figure 6: Cook's ship "Resolution"

Though he would have liked to continue to travel to distant lands, Banks settled in London, where for over forty years he organized voyages and sent botanists collecting roots, seeds, and living plants, if possible, from every far-flung part of the world—all destined for the king's gardens at Kew. Many of these voyages were to the coasts of North America.

One aspiring botanist Banks sent out, a young Archibald Menzies, returned seeds he collected around Halifax in 1784. He may have been one of the first Europeans to visit Nova Scotia who was more interested in botany than in timber and fish.

The Acadian French, by now, had been deported—ruthlessly we now admit—from their comfortable lands in the Annapolis Valley. Halifax was established and English-speaking settlers moved in to usurp the fertile lands of the Acadians. Settlers fanned out over the choicest parts of mainland Nova Scotia and Cape Breton, claiming lands with the best timber and agricultural potential.

Logging and clearing for agriculture—the double scourge of native plants—began immediately. We can only imagine the majestic forests and carpets of wildflowers we might have seen before it started.

Figure 7: Titus Smith

The first we know of to have had a good look was Titus Smith, a largely self-taught and unusually perceptive farmer, writer, geologist, geographer, botanist and surveyor. His talent was recognized in Halifax by "His Majesty's Governor", who, with his legislative assembly, desperately wanted a survey done of the interior of the province.

In 1801, they sent him off with these instructions:

> Your principal object...will be to visit the most unfrequented parts, particularly the banks and borders of the different rivers, lakes and swamps, and the richest uplands....You will make your remarks on the soil, the situation of the lands, and the species, quality and size of the timbers; the quantity of each sort also, and the facility with which it can be removed to market...and in every place which you shall deem calculated for these purposes, you will, as near as possible, estimate the quantity of acres, the possibility and means of rendering them fit for cultivation, either by banks, drains, or otherwise.

As usual, the emphasis was on the potential for logging and agriculture. Ironically, the crop deemed most promising for Nova Scotia in 1801 was hemp—needed to manufacture rope for the shipping in Halifax harbour. 222 years later it is the most promising crop again!

The mapping and survey work Titus Smith accomplished, under the most arduous of circumstances is admired to this day. With only a rudimentary map, a pencil and paper, and a compass, he spent over 150 days travelling through territory unknown except to a few trappers and natives, rained on, mired in bogs, lost in dense woods or clawing for miles through windfalls piled three feet high.

Despite the hardships—he apologized for his slow progress—his mapping and inventory of tree species were astoundingly accurate, and included a tally of native shrubs, sedges, grasses, ferns, and herbs.

In later life Smith remained passionate about native plants and their conservation, and in the 1830s he and painter Maria Morris Miller worked together to publish books of wildflowers. We now recognize that Titus Smith, an early natural-resources ecologist and advocate, was about a century ahead of his time in declaring that "progress" achieved by unchecked industrialization was just a scheme to create fortunes while destroying the lives of its human operatives and prospects for survival.

Figure 8: Mayflowers by Maria Morris Milller

The population of Nova Scotia continued to mount, with the burgeoning town of Halifax at its core. In the early 1800s, Dalhousie University and Saint Mary's University were established and a new breed of Nova Scotians began to appear. These were well-heeled and educated doctors, professors, scientists, clergy, and other professionals, a number of whom dabbled in botany.

Thomas C. Haliburton, the creator of the Sam Slick stories, authored a *Historical and Statistical Account of Nova Scotia* which included a section on native plants. For a time, though, there was no further progress on a comprehensive list, primarily because there was no established way to record discoveries.

This changed in 1863 with the publication of the first "Transactions of the Nova Scotian Institute of Science". Learned and semi-learned papers on botany and native plants were invited and published, and the list of known plants of the province grew rapidly.

Botanically-inclined citizens were quick to embrace the "'Transactions", contributing enthusiastic accounts of their discoveries as well as hints of their personalities and of societal attitudes of the day. Rev. E. H. Ball, who held down pastorates in Mulgrave, Halifax and other parts of Nova Sco-

Figure 9: The cover of 'Transactions'

tia, presented, in 1876, such an eloquent and impassioned paper on the subject of botany and ferns that I would almost like to copy the whole thing.

He writes:

> From its necessary tendency to call for walks and rambles into the country, in the woods and open fields, Botany is essentially a healthful study; and from the ardour with which it inspires its student, it gives an untiring interest.

And on the subject of ferns:

The Indigenous Ferns are graceful in habit of growth, they give charm to the landscape and have peculiarities of beauty and elegance which do not belong to flowering plants. Who has failed to notice the exquisite beauty of light and shade which towards sun-set characterize the small hillocks of *Dicksonia punctilobula* (hay-scented fern) so common generally along our road sides: how that the boldest dark shade is seen side by side with suddenly

Figure 10: Hay-scented fern

and almost imperceptibly blended lights, which, to an almost transparent whitish green, touch up the tips of the tufts when Sol's rays are nearly horizontal!

Wow!

In the same year, 1876, Prof. George Lawson of Dalhousie College, in the pages of the "Transactions", carried on a lively discussion with other botanists and various lay people on the question of whether English heather, Scotch broom, or the Great Laurel, *Rhododendron maximum,* which had all been found in Nova Scotia, could be considered native. Even the great Harvard University botanist Asa Gray was brought into the discussion.

After much back and forth, the conclusion was that broom was probably not native, but that heather and Great Laurel probably were.

Mr. Peter Jack, who reported to Prof. Lawson on the broom, and a man of his times, informed him that a large grove existed on the property of a "coloured" man, Mr. Jackman, near Shelburne. When in rich golden bloom it was a lovely sight. Mrs. Jackman "takes a great pride in the broom and is well pleased to show it to visitors, of whom there are several each year as its fame has gone abroad."

Figure 11: English heather

The English heather was not plentiful in 1876, but patches were found in several isolated and remote locations where it was un-likely to have been planted. Hence, it was believed to be native. Native heather was essentially identical to the European heather, and any discoveries of heather near towns and cities were suspec-ted to have arisen from seeds fallen from mattresses stuffed with heather or from heather brooms used by early Scottish immig-rants.

The final verdict that the exotic Great Laurel was indigenous rested on the fact that it was known to native hunters, who de-scribed "green bushes" with large white flowers. Because it kept its leaves in the winter, it was believed to be this species.

As proof that Great Laurel existed in the wild, collectors persuaded natives to dig up and bring them specimens, which was done in a couple of instances, until a fire apparently destroyed most of the grove and the rest died out—possibly because of collecting. The location of the site was forgotten and the plant has never been seen wild in Nova Scotia since.

Figure 12: Great laurel

By the 1880s much was already known about the native flora of the province. A.W.H. Lindsay. A professor at the Dalhousie School of Medicine, and evidently a formidable botanist as well, compiled an extensive list of species in his *Catalogue of the Flora of Nova Scotia* (1875-6).

Prof. George Lawson was back with *The School Fern Flora of Canada*, (1889), a text used extensively in schools in the era, when a knowledge of ferns was deemed more important to the young than it is today. Henry How, professor of chemistry and natural history at King's College in Windsor, submitted lists of plants observed in that area; George G. Campbell B.Sc., a list from the Truro area; G.H. Cox, plants from around Shelburne; and J. Fowler, a list from Canso.

Starting in 1883, Nova Scotia was favoured with repeated visits by John Macoun, a giant of Canadian botany. Macoun, an immigrant from Ireland as a young man, was a self-taught, self-made botanist.

Figure 13: Scotch broom

Hauling himself up by his own boot straps, he gave up trying to farm his claim of sour stony ground on the Canadian shield, bluffed his way into work as a school teacher, introduced himself to any botanist he encountered, boldly corresponded with even the most famous of them, worked tirelessly and turned himself into a respected botanist.

A chance meeting on a ferry with Sir Sanford Fleming resulted in an invitation to accompany an expedition across the Canadian prairies to the Pacific as a botanist. This expedition was primarily for the purpose of finding a possible route for a proposed railroad to unite one end of the country with the other.

The government in Ottawa wanted a botanist on the expedition to assess the suitability of the prairies for settlement, though in Ottawa, in 1872, the prevailing belief was that the prairies were a frozen wasteland suitable only for buffalo and nomadic Indians, and that British Columbia was probably not worth building a railroad to.

On the prairies, John Macoun quickly determined that belief was not so; that the land was extremely fertile and the hot summer weather perfect for crops. On his return to Ontario, he lectured on the advantages of a farm on the prairies and was almost single-handedly responsible for a great land rush of emigrants from the east.

With land on the prairies becoming quickly settled and large towns springing up, the government in Ottawa under Sir John A. MacDonald took the plunge to build a railroad across the country and through the mountains—a massive and almost impossible undertaking which, in the end, saved British Columbia for Canada.

Now, at the age of 52, after crossing the country more than once through uncharted wilderness, on foot, horseback and canoe, in the pursuit of new and undiscovered species—almost dying more times than you can count—Macoun wanted to have a look at Nova Scotia.

In the spring of 1883, he and his son William botanized around the Annapolis Valley, Cape Blomidon, Yarmouth, and Halifax, then hired a horse and cart and went all the way to Louisbourg, Cape Breton. In 1898 he returned to Cape Breton with his wife and youngest daughter. They stayed in and around Baddeck and botanized as far as North Cape. In Baddeck, the Macouns met inventor Alexander Graham Bell and found that "he and his family were most enjoyable people."

Figure 14: John Macoun

Macoun's single-minded passion for plant hunting and innocent disregard for what others might think of him were well known. While in Baddeck he recounts this amusing incident:

A government official, in conversation with the proprietor of the hotel in Baddeck, was overheard by Macoun's daughter to say, "If you had been here a little while ago you would have had a good chance to talk with an old tramp who was here and was all the time hunting round amongst the rocks looking for grasses and one thing or another."

My daughter, Nellie, said, 'That old tramp is my father!'

She tells me that the poor man nearly fainted.

Macoun remarks further that "We met a very many cultured people while at Baddeck and enjoyed ourselves very much and seemed to make a great many friends, some of whom are still friends of ours."

John Macoun's extensive collections from Nova Scotia are housed in Ottawa and recorded in his own life's work, the *Catalogue of Canadian Plants*, as well as in *The Flora of Nova Scotia*. The frequency with which his name shows up next to new entries in the Flora suggests that he didn't spend much time socializing at the Bells'.

With Macoun, the big man, come and gone, the part-timers kept things going. A. H. MacKay, Principal of Pictou Academy, published a list of new discoveries from all corners of the province. Some were new species, some were new sites, and others were mutant curiosities of common species like buttercup and dandelion. White variants of species that were normally coloured also caused much excitement.

A Dr. Somers, in 1885, reported on some new discoveries, including a white variant of "Low Ladies Slipper", submitted by Miss S. Gossip of Brunswick Street School, and an unusually large specimen of ground-nut from the garden of Mrs. W. Stairs. With wellbred gallantry, he commends these women for their interest, and adds, "It is cheering to us in our work to find ladies coming forward to aid in any department of it, and the least we can do is to encourage their good will, and endeavour to attach any who may feel inclined to the work of the Institute."

Also from the Pictou Academy, C.B. Robinson B.A. observes, and rightly so, that, considering the increased attention to the teaching of native plants in schools, it is surprising that the early-blooming intervale flora, found along the rivers and streams of Eastern Nova Scotia, is virtually passed over. The intervale flora, which occupied the banks of most significant streams of this region, included many of the choicest and most beautiful species in the province. Maybe they were overlooked because they bloom early, before the over-hanging trees leaf out in spring, and are easily missed. Because they bloom early, and often die down soon afterwards, they are, in fact, termed spring ephemerals.

Robinson mentions specifically the bloodroot, the trout lily, and the Dutchman's breeches, but includes the trilliums, yellow violet, Spring beauty, toothwort, dwarf ginseng and, queen of them all though not a true ephemeral, the Canada lily.

Figure 15: "a garden of irises"

In 1905, Nova Scotia could be a wild place, and, aware of this, Walter H. Prest of Bedford published an extensive list of edible wild plants of Nova Scotia based on his experiences in the backwoods, to be of help to anyone who might find himself lost in the woods and hungry. He writes:

While some of the wild fruits here mentioned, such as the blueberry and cranberry, are of commercial value," he says, "others are included because they may assist in sustaining life at a critical time. While lost in the forest, persons have perished through a want of knowledge of the resources that nature has bounteously provided in many sections at certain seasons of the year.

Also in 1905, Captain John H. Barbour, M.D., Royal Army Medical Corps, and obviously a plant lover, comments on the flora of McNab's Island in Halifax Harbour. Describing the profusion of flowers, he includes the rock roses and masses of purple iris so prolific that the island could be called "a garden of irises." He notes as well, the tremendous quantities of wild raspberries. A good section in which to be lost, I think Mr. Prest would say!

Captain Barbour shows a practical turn of mind and wonders about the feasibility of harvesting raspberries for the Halifax market. His practical mind also suspects that there could be a commercial use for the irises. "Would it not be possible," he wonders, "to manufacture a cheap and beautiful violet ink, stain, or dye, from their rich, velvet perianths?" Watch out, wildflowers!

Figure 16: Blue-eyed grass

In another article, though, Barbour vindicates himself with an innocent treatise on the little relative of the iris, the blue-eyed grass. He admits that he knows very little about Nova Scotian flowers because he has only been in the country one season, but would like to report that he has observed great variation in specimens of the blue-eyed grass—a species he believes to be as common here as is the primrose in England.

At the turn of the century, plant exploration began, in Nova Scotia, to take on an entirely different aspect. With the appearance, in 1899, of the first edition of *Rhodora*, the journal of the New England Botanical Club, botanists from Atlantic Canada and from New England began to work together, publishing their findings, for the most part, in *Rhodora* or the *Transactions of the Nova Scotian Institute of Science*.

With the publication of the seventh edition of *Gray's Manual of Botany* by the renowned botanist Asa Gray of Harvard University in 1907, botanical exploration in Nova Scotia entered a new and decidedly more professional phase. Though it was surmised that Nova Scotia had, by now, been thoroughly botanized, with little left to discover, some were not so sure. Enter the big guns—dedicated and fearsomely competent career botanists from the prestigious universities of New England, ready to check up on the Canadians.

The rugged valleys, crags, and highlands of Cape Breton were the first to attract attention. Professor George E. Nichols of Yale University was an avid outdoorsman, an ecologist and a botanist. He made repeated visits to Nova Scotia, beginning in 1905, camping and exploring, and doing some botany on the side.

Figure 17: The valleys, crags, and highlands of Cape Breton

Stating that, to his knowledge, only three other botanists (John Ma-coun, C.B. Robinson, and J.R. Churchill) had done any plant collect-ing in Cape Breton, Nichols began an ecological study of the veget-ation of Cape Breton in 1913. The study took the better part of four summers and an impressive amount of outback camping and travel on foot throughout the island.

In the summer of 1916, Nichols and a companion traversed the entire shoreline from St. Ann's Bay to Cape North on foot, then spent a week in the interior. Nichols' work in Cape Breton was thorough, and an important record of the ecology and the general flora of northeastern North America at the time.

By now, serious botanists were no longer primarily interested in undiscovered species. It had become quite uncommon in the Northeast of North America to discover a new species.

What interested botanists about Cape Breton were discoveries of species known elsewhere but not expected to be found in Nova Scotia. These were certain arctic-alpine and boreal species that had migrated north following the glaciers as they receded 10,000 years ago. The usual range of these species by now was far to the north in Labrador, northern Quebec and the Gaspe, where they had been driven by warming temperatures and aggressive herbaceous competition from the south.

These rather fragile, cold-loving species were long gone from most of Nova Scotia, but populations were being discovered in shaded ravines and cliffs, beside waterfalls, and on the highland plateau of Cape Breton. These populations, growing far south from where the species should be found, were called disjunct popula-tions and disjunct species.

The study of these disjunct populations, beyond being interest-ing in itself, could provide important clues to the course of the last glaciations in the Pleistocene and the timing and paths of the plant migrations that followed the glaciers coming and going. This was the magnet that attracted the American researchers to the province.

After Nichols came a procession of botanists, connected with the Gray Herbarium of Harvard University, to investigate those disjunct species in Nova Scotia of which there had been rumours. The Gray Herbarium, founded in 1843 by Asa Gray, the greatest American botanist of the nineteenth century and friend of Charles Darwin, had become one of the most prestigious in the world. There was no important herbarium in Nova Scotia at the time and specimens collected in the province were sent to the Gray, or to the national herbarium in Ottawa.

A herbarium was then, and still is today, a museum of sorts where pressed and dried plant specimens are catalogued and stored for research.

Nobody has found a better way to preserve collected plants than by drying them. They are pressed to flatten them and save space. Properly prepared specimens can still be good after a hundred years or longer.

The procedure was worked out centuries ago and involves layering the carefully arranged plants between sheets of absorbent paper and pressing them flat under pressure in a press. The ar-

Figure 18: Herbarium sheets

rangement in the press is of prime importance because, after drying and flattening, the flower parts must still be identifiable and allow examination. A well-dried specimen stored in the dark might even retain the colour of its petals. Also, there will be a piece of stem showing the size and arrangement of the leaves, and often roots. too.

The press is placed in the sun or a dry room somewhere until the specimens are completely dry. Then the nicely flattened plants are taken out, mounted on stiff paper, layered in boxes with paper in between, and sent off to the experts at a herbarium. If the specimen is properly arranged on the sheet, the experts have all they need to determine if it is actually the species it is supposed to be, or maybe even a new one.

Many of us were hooked on botany at university after being required to collect, dry and mount 50 or so different species of plants for our plant taxonomy class. The first 25 species were easy to find, but after that it required a lot of combing through interesting, out-of-the-way places for new ones. The search became very exciting, and the pressed and mounted specimens were good-looking enough to be put up on the walls of our dorm rooms.

You might like to try some yourself. Even cultivated flowers around the house can be pressed, dried and mounted. There are books, and good instructions on the internet.

So, back to the 1920s. The preparation of properly dried, mounted and labelled specimens was essential for every plant collecting expedition. It was critical that these were carefully done and suitable for examination at a later date in the laboratory. Collecting plants was the easy part. Drying them could be hell.

Merritt L. Fernald spent the summers of 1921 and 1922 collecting specimens, mostly in southwestern Nova Scotia, leading a team of botanists and students of botany from Harvard. He had determined that this was a region of myriad lakes and vast swamps and bogs that botanists had scarcely touched. In addition, he had heard that some intriguing disjunct species, of which I will write later, were hiding there.

Fernald and his entourage did not come unprepared. He writes:

> Altogether there were eight in the party, though not all at one time. 5000 sheets of drying paper, nearly as many cor-

rugated "ventilators," a large stock of white pressing paper, seven large collecting boxes, ten presses, a bushel of flake naphthaline (to keep out mold and hasten drying of "soggy" specimens) and the other necessary equipment (to the extent of 16 heavy freight boxes)...

Figure 19: Merritt Fernald (l) collecting in Virginia in 1942

However, Fernald and his group, with headquarters in Yarmouth, were about to discover the reality of southwestern Nova Scotia.

The reality was fog. Or rain. Or fog and rain. It was completely impossible to dry plants outdoors.

The wet weather became a joke. Planning to explore around a promising lake at the end of the day, Fernald remembers:

> Approaching sunset warned us before we had got half the length of the west shore that our plan to encircle the lake was too ambitious. The fog was still with us and during the eight-mile road-walk into Yarmouth we amused ourselves vainly attempting to make out the outlines of more than two of the roadside telephone poles at a time,—an index to

the extreme density of the atmosphere.

It was some days after this, when the uninterrupted fog was in its fourth week, that Mrs. Graves wrote home that they had been there for a week but had not seen Yarmouth yet.

Fernald also comments on the peculiar scarcity of mosquitoes in the area:

> We had been most happily surprised to find that we could go anywhere on these boggy barrens without meeting this much-to-be-expected tenant. But in explanation some one suggested during the summer, that in such a dense atmosphere mosquitoes, if they there exist, must remain in the larval stage, wings being quite useless to them!

All joking aside, the rain and fog were a serious impediment to the all-important work of drying and pressing specimens. The group were able to use a barn to get out of the rain, and some American ingenuity:

> We reached Yarmouth that evening and the next three days were occupied until late in the evenings with our presses. The 5000 driers proved wholly inadequate, for Yarmouth was wrapped in its conventional blanket of fog and sun-drying was out of the question. We had already been driven to various expedients to meet the penetrating dampness and now with great regularity, as soon as corrugated ventilators had been inserted, the presses were stacked high in a square about the kerosene stove or suspended over it from the rafters. The wet driers for immediate use had to be "toasted" while such as could be allowed a more prolonged aeration were tucked end-on into chinks in the rough boarding of the empty hay-loft. The act of thus fitting the rough ends of the driers into shallow chinks from which they drooped soon became a real art and with the aid of a ladder we were eventually able thus to decorate the rough

sloping walls of the loft with nearly 2000 driers at one turn.

With a good crew, and this sort of savoir-faire, Fernald was able to report at the end of his two years in Nova Scotia, 17,000 well prepared herbarium specimens including many completely new discoveries for the province. And he also proved his hypothesis.

Though Merritt Fernald and his group contributed enormously to what was then known of the flora of this province, and discovered species previously unknown to occur here, the discovery of new species was only a sideshow. Fernald was looking for species he knew well, which were commonly found along the Atlantic coastal plain from New Jersey to Cape Cod, but unknown farther north.

He believed that in the Pleistocene period of the earth's history, the oceans were much lower than today and that there existed a land bridge connecting what is now Nova Scotia with the southern coastal plains of what became the U.S.A. 10,000 years later.

The theory goes that plant species pushed southward by repeated glaciation were able to migrate again northward as the glaciers retreated, staying well out on the broad low-lying land skirting the rocky coasts of what are now Massachusetts and Maine until they arrived on our shores. With the melting glaciers, sea levels rose, cutting off the bridge over which plants had migrated, leaving individuals stranded at either end of the route.

If Fernald could find certain key species in Nova Scotia which were found also on the southern coastal plain of the U.S.A. and nowhere in between, as he had reason to believe that he would, he would prove his theory.

These were another sort of disjunct species, and he found them. There were inkberry, golden crest, meadow beauty, Plymouth gentian, a slew of sedges and grasses that I've never heard of, and the star of the show, curly-grass fern.

The curly-grass was an unassuming little fern resembling a grass, which was, in fact, often concealed among true grasses and difficult to spot. It was native in damp savannas among the pine barrens of New Jersey; a single specimen was reported from Long

Island, but nothing further north. Surprisingly to many, but not to Fernald, it was said to have been found in Nova Scotia—far north of any other known population. It could only have come on the land bridge.

Over the next two summers, Fernald and his crew found substantial populations of the disjunct species, including curly-grass fern. They returned to Harvard convinced that the existence of a prehistoric land bridge could no longer be disputed. In his own words:

> Further exploration will greatly increase the proportion of isolated coastal plain types, for we have glimpsed scarcely 1% of the silicious area and most of the significant plants are highly localized and found where least expected. But if there were need of further evidence that, since the Pleistocene glaciation, the continental shelf of eastern North America has been high in the air, affording an essentially continuous line of migration across the mouth of the Gulf of Maine to Nova Scotia, thence to Newfoundland, that evidence is now abundantly at hand.

Figure 20: Curly-grass fern

Why significant plants were highly localized and found where least expected puzzled professor Fernald, and he didn't like to be puzzled. He observed that species in Nova Scotia often grew in drier and more exposed locations than did the same species in New Jersey. Also, there was the question of how these same species were able to survive in a so much colder environment. As he was wont to do, he thought it out thoroughly, and was forced to admit that it was no doubt a side effect of that good-for-nothing fog and humidity he had endured for two summers.

The plant list for Nova Scotia was growing rapidly. Researchers continued to visit from Harvard, and took an interest in offshore islands.

In 1921, Harold St. John studied the vegetation of Sable Island. Despite the fact that Sable Island has the mildest winter temperatures in the province, St. John found the number of species rather limited due to the extreme exposure of the island to winter blasts off the Atlantic.

In 1931, Lily M. Perry, a botanist trained at Acadia University and working from Harvard's Arnold Arboretum, spent the summer camping and collecting with Muriel Roscoe, also from Acadia, on the windswept St. Paul Island off the northern tip of Cape Breton. Collecting over the tortured land-

Figure 21: St. Paul Island

scape of St. Paul was tough, but the women maintained their sense of humour. Lily writes:

> Owing to zones of dense undergrowth of brushwood, travelling in the forest was exceedingly difficult, and in many places well-nigh impossible. Sometimes the scrub was sufficiently compact to carry our weight, but oftener it swallowed us and we had to hunt for a way through.

After a hard day's exploring, the women changed direction.

Lily says:

> Lena Lake was a different proposition. We had been warned
> to keep away from the lower end if we valued our lives. The
> muddy bottom, although seemingly solid, had several times
> given way suddenly, so that it was impossible to ascertain
> the real depth of water. But alas, here if anywhere ought to
> be good collecting. The vegetation grew from the margin
> nearly to the middle of the lake. We visited the place soon
> after landing and noted that the plants were too immature
> to collect; incidentally, we discovered by stepping into the
> water, that good-sized leeches were there in abundance.

As did Fernald, they found the fog and humidity made drying
plants very difficult. Nevertheless, they accomplished a prodigious
body of work, managing to dry and mount 2360 sheets of specimens of which twenty were brand new discoveries for Nova Scotia.

They also were curious about what the species collected on St.
Paul Island might reveal about the pattern of prehistoric glaciation.
Of the twenty species new to the province, Perry found that most
of them were normally found farther north. She explains:

> It is significant that these northern plants are characteristic
> of slightly or not at all glaciated regions in Newfoundland,
> Quebec, Anticosti or the Magdalen Islands; hence, if I may
> draw a parallel conclusion, it seems not unreasonable to in-
> fer that St. Paul also escaped denudation by the Wisconsin
> ice-sheet.

So another piece of the puzzle and another twenty species for the
Flora.

Another highly-respected woman botanist was working at this
time, Margaret Sibella Brown from Sydney Mines, Cape Breton. The
daughter of Richard Brown, general manager of the Cape Breton
coal mines and first elected mayor of Sydney Mines, she was educated in Germany and London, England, receiving training in china

painting and French.

Back in Nova Scotia in 1885 at the age of 19, she studied at the Victoria School of Art and Design. By 1934, she still worked in art, and was awarded an honorary diploma for helping in the transition of the Victoria School to the newly-established Nova Scotia College of Art and Design.

By this time, she had established, as well, an international reputation as a field botanist specializing in mosses and liverworts. She trained and travelled the world with experts in the field and published numerous papers on the subject. She never received a university degree but her scientific qualifications were beyond question.

Her field work inevitably led to discoveries of plants other than mosses and liverworts. In 1939 she reports in the *Transactions of*

Figure 22: Margaret Sibella Brown

the Nova Scotian Institute of Science on her discovery of a new species of bladderwort, *Utricularia inflata*, for the province. At 73, her joy of discovery and delight in nature still are evident:

> Specimens, in full bloom, of this rare plant were obtained in this lake (Sawlor) from two distinct places, floating among fragrant white water lilies, yellow cow lilies, tiny floating hearts and colonies of pickerel weed, all in a cosy, sheltered cove circled by the black reflections, upon the quiet waters, of spruce and hemlock trees lining the lake shore.

Margaret was elected president of the Halifax Floral Society and, at the time of her death at age 95, she was the oldest living member of the Nova Scotian Institute of Science.

The last of the big guns from Harvard to visit Nova Scotia in this era was Charles Weatherby in 1942. He came to Nova Scotia in search of a rare species of plant, the Redroot, just because he wanted to see it—proof of an avid botanist.

Weatherby was not coming to rough it, like some others. Physically, he was not robust as a result of illness in youth, but this had no bearing on his botanical ardour. In fact, it was why he chose the profession:

Figure 23: Charles Weatherby (l) on a collecting expedition

I was an invalid for five years, twice given up by my physician, and was never really rugged thereafter. From this experience I learned that literature, my chief study in college, was for me a poor support *in extremis* and that science (I had botanized as a hobby) offered a much firmer foot-hold.

Weatherby and his wife, Una, also a botanist and an artist, travelled by car, searching for promising sites near the road. In the evenings they were established comfortably in Mill Village on the lower Medway river, Queens County.

We chose our headquarters happily; our landlady, Mrs.F. Laurie Mack, not only provided us with comfortable lodgings, well suited for botanical work, and excellent food, but was interested in our activities and most helpful in securing for us needed information.

In the daytime, they hit the jackpot when they drove up the Medway to Ponhook Lake, where they found the Redroot, as well as many other species, some unknown in Nova Scotia and even a few unknown in Canada.

In his account of this excursion to the province, Weatherby makes reference to a herbarium at the Nova Scotia Department of Agriculture in Truro, which must have just been started, and thanks Mr. A.E. Roland, the Provincial Botanist, for his help—the same Mr. A.E. Roland who will play the next leading role in this history.

In 1946, four years after meeting Weatherby, Albert E. Roland published the first edition of the *Flora of Nova Scotia*, an ambitious compendium of all the plant species known in Nova Scotia at the time. This was something which had not been attempted since Lindsay's flora seventy years before.

Roland began his Flora in 1944 while a graduate student at the University of Wisconsin. Upon graduation, back in his native Nova Scotia, he was appointed Provincial Botanist. He also taught at the N.S. Agricultural College in Truro, and worked on the book.

Figure 24: Nova Scotia Agricultural College

Dr. Roland, in addition to teaching and inspiring hundreds of students at the A.C., was an avid plant collector himself, discovering species new to Nova Scotia and adding them to the growing herbarium at the College, of which he was curator. He worked tirelessly in the herbarium, examining sheets of new and old discoveries, adding entries, sorting out classifications and working out the "key" for a future edition of the Flora.

After 1947, says Roland, the plants of Nova Scotia were again intensively studied, with emphasis upon Northern Cape Breton, which was still relatively unexplored due to its rugged and inaccessible nature. It was an eventful time.

Merritt Fernald, back at the prestigious Gray Herbarium at Harvard, published his own masterwork, the 8th edition of the *Gray's Manual of Botany*. Both Roland at the Agricultural College and E.C. Smith, a professor at Acadia University led summer field parties, sponsored by the Nova Scotia Research Foundation, into the wilds. Plants discovered by these field parties were dried and mounted and an important herbarium collection was underway at Acadia. There were significant herbaria also at the Agricultural College and at the Nova Scotia Museum of Science in Halifax.

By the 1960s, the discovery of a new species was becoming a fairly uncommon occurrence, but there was plenty of habitat yet to explore. The plants of the more accessible forests and river intervales of the province, at least what was left of the forests and intervales, were well known. There were still, however, vast expanses of bogs and barrens, waterfalls, cliffs, and rugged gypsum outcrops largely untouched.

Fernald and his team of botanists, for instance, after two summers of intensive study of lakes, bogs, and barrens in southwestern Nova Scotia, reckoned they had surveyed less than 1%, and much of that only superficially. No one had really been there since.

It was the same story everywhere in the province where the land was steep or wet or otherwise inaccessible—just the sort of places that intrigue the botanist.

John S. Erskine was a teacher in the Annapolis Valley, and a disciple of Margaret S. Brown. His main focus botanically were the liverworts and mosses, but he was a skilled all-around botanist. He had a passion for the archaeology of Nova Scotia's native people, and outdoor botanizing.

In 1957, he published a study of the Tusket Islands, a foggy, bleak and windswept cluster off the mouth of the Bay of Fundy. In this study, he considers the species and vegetation patterns on the islands and how they are determined and affected by the activities and sporadic incursions of the island's poor and largely transient population of farmers, fishers and herders.

Figure 25: The Tusket Islands

On the weekends, John Erskine, it seems, was off to the marshes, barrens, bogs and gypsum cliffs with his wife, hunting new plant species, or exploring an ancient aboriginal encampment. He published articles on his discoveries and theories, and contributed significantly to the list of known plant species of the province.

The wilds of the Cape Breton highlands, and the lure of rare species hidden in the cliffs and chasms through which its rivers flow, were still a powerful magnet for researchers. Macoun, Nichols, and Fernald had done what they could, but had hardly scratched the surface. The search was still on for elusive survivors, spared by the glaciers but stranded in steep, shaded or otherwise inhospitable refugia, safe from more aggressive species.

R.W Hounsell, W.B. Schofield, and E.C. Smith, botanists from Acadia University in Wolfville, all heeded the call.

E. Chalmers Smith was a biology professor teaching plant identification, taxonomy and ecology. His teaching and enthusiasm launched many an eager young botanist. He became a legendary figure at Acadia, serving as Vice-President Academic for over 25 years. He authored or collaborated with others on numerous botanical publications, and collected, with his students, thousands of specimens for the herbarium at Acadia, eventually named the E.C. Smith Herbarium in his honour. He was, in addition, a good-hearted man.

Prof. Smith was born and raised in Hillsboro, Cape Breton and, like most Cape Bretoners, was loyal and, though Wolfville wasn't far, knew what it was like to live away. Each winter at Acadia, he and his wife hosted a supper for any homesick young Cape Bretoners in the department that year. In 1973, my wife, Mary, was one of those, and still remembers the kindness the Smiths showed to their guests

In 1968, E.C. Smith and his junior researcher, R.W. Hounsell, made an ambitious study of six steep river gorges in the Cape Breton highlands, focusing on waterfalls, wet cliffs, and talus slopes where alpine and boreal disjunct species were likely to be found. They did find disjunct species at each location—some locations more rewarding than others:

> Practically every river valley examined to date has been shown to possess one or more disjunct stations, none of which is very extensive in size; some are a few hundred yards long, others much smaller. Some provide an acceptable niche for only one species, but most harbor several.

E.C. Smith continued to teach and manage the herbarium, and, in 1969, published, with Albert Roland, the long-awaited *Roland's Flora of Nova Scotia*, 2nd edition.

So there you have it: Roland's Flora. A triumph indeed. The culmination of exploration, discovery and research by hundreds of individuals dating back nearly 200 years. A book that was twice as long in the making as the bible, but which, tragically, was obsolete the moment it was printed.

The best a printed Flora can do is catch up to the living one. The living flora never sleeps. New species come and go, species distribution changes, and taxonomists rearrange entire families. Overnight, the printed Flora is not quite right, and to catch up again will require a new one.

Figure 26: The Flora

Meanwhile, the seasons went round and round. The collections at the E.C. Smith Herbarium increased to nearly 200,000 specimens, under the stewardship of Dr. Sam Vander Kloet, Ruth Newell, and Dr. Rodger Evans.

Botanical exploration and research began to change. The field botanists were less interested in making plant lists, and more interested in monitoring populations of rare and endangered species. Botanical researchers published more articles on discoveries in genetics and morphology than they did on discoveries in the wild.

Exciting discoveries were still made from time to time, nevertheless, and probably always will be. Four examples: a new orchid, the Southern twayblade, discovered in a bog in Inverness County, N.S., in 1969; a patch of the large-flowered white trillium—the provincial flower of Ontario—discovered near Centreville, Kings County in 1976; A new site for the Maidenhair fern on the Meander River intervale near Brooklyn in 1989; and, finally, also in 1989, the discovery of *Toxicodendron vernix* around the inflow to Telfer Lake in Queens County. Nice to know we have poison sumac in Nova Scotia, isn't it!

Interest in the Cape Breton Highlands and disjunct species remained strong, with Harold Hines discovering several new species, including a new dwarf rhododendron—the Lapland Rosebay—in 1984, in the Corney Brook gorge. By 1993, Rene J. Belland and W.B. Schofield were still discovering rare species in the gorge, including a species of willow and a saxifrage new to Nova Scotia.

Going into the 90s with plant exploration in full swing, and ongoing re-examination of species in the laboratory, Nova Scotia was overdue for a new edition of the *Flora*. Albert Roland, who never stopped, was well along towards a third edition when he was forced, because of his health, to give it up in 1990. Marian Zinck took it on.

The third edition of *Roland's Flora of Nova Scotia*, after many delays, was released to an eager public in 1998. It had been a massive undertaking. The new two-volume format with modernized keys, enhanced illustrations, updated maps and distribution data, pronunciation guidelines, etc., required ten years of intensive work by Marian and her team of botanists, artists, and editors, and is as much her flora as it is Roland's.

Ironically, though a wonderful piece of scholarship, it, too, was obsolete immediately—the inescapable fate of any printed Flora. Sadly, another print edition was out of the question. Digital would be the way to go.

Marian herself commented:

> I knew soon after the publication of *Roland's Flora of Nova Scotia* (1998), that an eBook was inevitable. Aside from the additional species records gained from countless hours spent in the field, publishing technology was evolving rapidly, embracing digital delivery.

Marian, along with Ruth E. Newell and Nicholas M. Hill, have embraced that digital delivery, and created, in fact, an e-book with photos of practically every species. Better yet, it is available to the public as a free download. Perhaps it will never become obsolete. It can be added to and updated constantly.

The Flora is a living thing. It is not a book that sits on the shelf and gathers dust. No matter which printed edition you own, or the eBook, it is often out of the bookcase and open on the table. If you are out collecting plants, there is always something to look up.

My wife and I still favour the little green hard-cover '69. This, the second edition, still has almost everything in it. It is easier to slip into the day pack or read in bed than is the two-volume set.

We have two copies of the second edition—one at home and one at the cabin—and the two-volume third edition on the bookshelf. At our place, Roland's tome gathers no moss!

Figure 27: Bloodroot and trout lily, floodplain companions

2: The Floodplain

Figure 28: The floodplain in early spring

With warm weather finally on its way, it is time to dream of wild places and native wildflowers. Any flowers at all, for that matter, at this time of year.

The concept of "native" is very nebulous. Visitors from other lands have been coming to Nova Scotia since the 1600s, bringing along their favourite flowers from the old country. Our most familiar weeds came with their grain, or the ballast in their ships, or maybe the mud on their boots. All of these have had plenty of time to spread around the province and "go native."

Often it is unclear whether or not the same species of plant was here already. I think that professional botanists can sort this out by counting chromosomes, but we amateurs can't. We just learn their names from the field guide and look them up in the Flora when we get home.

A foray into the great outdoors to check out the plants is never a waste of time. Once you get interested in plants you can never be bored.

If you can't make it to the wilderness, you can learn a lot just walking around your town or city. You are sure to see beautiful trees and flowering shrubs, and flowers in the flower beds. By all means find out what these are. They almost all have counterparts in the wild.

You may also be surprised by how many species of plants grow in vacant lots or cracks in the sidewalk. The common milkweed is one example that has risen to celebrity status recently as a host for the monarch butterfly.

Of course, there are the public parks, too, and more and more areas near cities are set aside as public wildlands. Some of these feature good populations of native wildflowers. If you are visiting another province or another country, stroll around and see what grows there. This is an ideal way to spend your time and save your money.

The province of Nova Scotia, considering its comparatively small land mass, offers to both plants and plant explorers an impressive variety of habitats. Within human memory, which spans but the blink of an eye in geologic time, Nova Scotia has never been a notably rich agricultural province. Scraped down time after time to bare rock by the glaciers of prehistoric times, its soil is thin and unevenly distributed.

Granted, there are areas, mostly along river valleys, where the glaciers left us some soil, and a modicum of agriculture is possible. A bit shortchanged on agricultural land, maybe, but we are a province rich in bogs, barrens, swampy wetlands, seashores, cliffs, and stony uplands, and for this we should be thankful. This is where our wildflowers hide.

Places to look are wherever agriculture and forestry are diffi-cult, and the landscape is therefore largely untouched. The essentially rugged and poorly drained aspect of much of Nova Scotia leaves us plenty of possibilities.

As the snow melts in earliest Spring, though, rugged areas can wait. We visit instead what is left of our gentle river valleys and hardwood uplands. Here, at this early season, bloom the delicate wildflowers that once covered so much of the province.

The Floodplain

The major river valleys of Nova Scotia are neither rugged nor inaccessible, and boast the richest soils of the province. At the time of colonization, this is where the most majestic forests were found, and undoubtedly the most luxuriant carpets of wildflowers. These flat, fertile river intervale lands, or floodplains, were the first seized by colonists for clearing and agriculture.

So, what happened to the wildflowers, and why do we start looking here?

In short, the wildflowers were almost exterminated, but not completely. Fragments, giving a glimpse of what must once have existed, can still be found where land was not cleared all the way to the river bank. Clearing the trees is a death sentence for the wildflowers that depend upon the shelter and the shade and the rich soil.

Our native wildflowers are not weeds. When the trees are cut, they can't simply move and grow somewhere else. They wither and die and will never be seen again in our lifetime, even if the trees grow back.

On this subject, my friend Henri, when he was a student at the Agricultural College in Truro, went to talk to the professor, Albert Roland himself, about the plight of native trees in the province. At the time, the spruce budworm, the beech bark disease and the Dutch elm disease were wreaking havoc in the forests, and clear-cutting was rampant.

Dr. Roland's response was to not worry about the trees, they would grow back. Worry instead about the little plants. They will vanish and be gone forever.

The little plants that flower beneath the leafless hardwoods of river intervals and rich uplands are called spring ephemerals, so named for their fleeting bloom. To my mind, these are the most precious of our native wildflowers, blooming briefly but in profusion, shortly after the snow melts in Spring. They choose their days.

Weather at this time of year, late April or early May, is capricious. Don't look for them on a cold and windy day. They stay folded up and hidden until that lovely, mild, sunny day when the birds are singing and calling us outdoors. Then they rise and open wide to add their glory to the day.

The list of players in this Spring show starts with bloodroot, Dutchman's breeches, trout lily, spring beauty and trilliums. They have evolved to shoot up quickly in the sun, bloom, and complete their reproductive cycle before the trees overhead leaf out.

Figure 29: Bloodroot

Bloodroot, usually the most conspicuous of them all, only opens its flowers when the sun shines, and the beautiful white flowers with the yellow centres blow apart in a day or two. Bloodroot is easy to miss.

The other ephemerals don't bloom for much longer. That is why we are talking about these wildflowers first. They are the earliest to bloom and the first ones to look for.

The Spring ephemerals along the riverbanks of our rivers and streams are absolutely dependent upon the trees overhead that will shade and shelter them when they are finished blooming. Sometimes they may be found beneath choke cherries or alders, which will do the same thing.

Fortunately, some landowners didn't clear their land to the very edge of the stream, and may have left a grove of old hardwoods. As a matter of fact, they are now required by law to do so.

In Roland's Flora, a river intervale at Kemptown, near Truro, is repeatedly mentioned as having the best intact population of rich floodplain flora in the province. Spring beauty, bloodroot, trilliums, trout lily, wild leek, yellow violet, blue cohosh, foamflower, and even the stately Canada lily grew together under stately sugar maples.

Figure 30: Trout lily on the riverbank

For years I hungered to visit this place. I imagined a vast, hidden intervale somewhere back in the Cobequids where botanists wandered knee-deep in flowers.

51

One day I made it to Kemptown. I wasn't sure I was in the right place. All those flowers were there, yes, in an intervale about the size of a tennis court. The stream was rather small and crossed by a bridge. The stately sugar maples were long gone. The flowers were hemmed in between a cow pasture and the stream, and grew mostly under the chokecherries and alders. On the other side of the bridge was a modern bungalow and a camp down by the water.

A couple of years ago I visited again. The cow pasture was bigger and the intervale smaller.

On May 20, 2022 I went again with Mary. The place is still interesting as a sort of small museum of formerly-abundant intervale species, but is steadily shrinking.

Uncommon in our end of the province, this is the only place we have seen the yellow-flowered trout lily, named for its mottled glossy leaves which resemble the back of a speckled trout. Surprisingly, the lily and other of the smaller wildflowers in this intervale now seem to be getting swallowed up by the big, straplike leaves of the wild leek, which is itself generally rare; and we didn't find foamflower, spring beauty, or Canada lily.

Things are always evolving, and we'll hope to see them next Spring.

We did return May 14, 2023. I would say things are looking up. We were there a week earlier than last year, and there was much more trout lily than we had been aware of, and more bloodroot. Also, the cow pasture seems to have been abandoned and is growing up again in trees. This may be a big wildflower garden again someday!

Mary and I have found our own sites along rivers here in Antigonish County. Early in Spring 2022, we headed for a couple of our favourite spots to see if they were still as good as we remembered.

Our best spot is a stretch of about a kilometre of old, somewhat-decrepit forest between a large hay field and a smallish river. This strip is anywhere between 50 to 400 meters wide along its length. A quite respectable list of the spring ephemerals are found here. We have been visiting this intervale for over thirty years, often going for fiddlehead ferns, and it has remained remarkably undisturbed. By man, at least. The elements are a different story.

The forest consists mainly of ancient, widely-spaced maple and ash, some rotted, some hollow, some fallen, and some standing with woodpecker and nest holes. Sound trees still standing are enough to provide shade, though many are broken and battered by ice and wind storms. Broken branches and limbs and rotting logs litter the ground.

In flood conditions, the river overflows its banks across the floodplain, ridding itself of silt and debris that would otherwise clog the channel. Here there is evidence of serious flooding—windrows of leaves and river grass washed up in the fallen branches, and expanses of bare mud. Just the place to find wildflowers.

April 28: Our first visit to the riverbank. Weak sun and cloud. 7 degrees. Signs of spring are ramping up. Alder catkins are dangling, and yellow with pollen. Pussy willows are out and the red osier dogwood twigs are flaming red with rising sap. Bird song is everywhere, and we hear our first motorcycle.

Walking down to the river, we pass quantities of coltsfoot

Figure 31: Coltsfoot

blooming alongside the path. It now has been blooming for weeks.

Is coltsfoot a native species? I think maybe the jury is still out on that one. At any rate, it puts on a lovely show on a sunny day, and pops up in the worst gravel and clay. Its yellow dandelion-like flowers appear before the leaves as the snow is melting andwe are desperate to see something in bloom. If it were rare we would all be out searching for it.

At the river bank, the true ephemerals are not quite as far along, but we see bloodroot poking up through the mud and starting its flower buds. Bloodroot comes on quick and might be blooming to-morrow. There is no evidence yet of nodding trillium, yellow violet, or Dutchman's breeches, but we expect to see them soon.

Figure 32: Yellow violet

We see the clumps of ostrich (fiddlehead) fern that we knew would be there, but they are still asleep. Many seem to have disappeared beneath the heavy layer of mud left by winter floods.

This mud covers large areas and is deep in places. I cringe to use the word 'mud', but I can think of no other. Imagine instead a life-giving layer of nutrient rich, alluvial clay and silt scoured from the flooded stream bed, replenishing soil fertility and feeding the wild-flowers. In the process river rocks and gravel are washed and cleaned, so trout may spawn.

Where uninterrupted by man-made channels and dams, this natural cycle takes place on nearly every river system in the world, and is the source of soil fertility that feeds humanity. Think of the Nile.

Next day, April 29, there is some bloodroot in full bloom. It is a dull day, though, and most flowers are only partly open, showing the purplish colour underneath the petals.

The bloodroot is one of those flowers that close up for the night and only open wide on sunny days. Not only do the flowers close up at night, but the leaves rise up and wrap around them until morning. In the warm sun, though, the hundreds of flowers wide

Figure 33: Dutchman's breeches

open above the broad leaves, with their snow-white petals and yellow centres, are a sight to see.

Meanwhile, there is evidence of Dutchman's breeches and trillium starting to come up, and yellow violet in full bloom. I like this violet, maybe because it is uncommon and is found in rich woods early in Spring. It is tall, with bright-looking leaves and small cheerful flowers.

As a matter of fact, I like all the violets. There is no such thing as a bad one. We can look forward to handsome clumps of the common purple violet in sunnier locations, and the tiny, sweet-smelling, though unimaginatively named small white violet in wet spots. There are no less than 17 species of violet listed in the Flora, all lovely and unpretentious.

As soon as the sun comes out, we can be assured that the blood-root will put on a good show—for a couple of days. Almost right away the petals start to blow off so seed pods can form. The very embodiment of a spring ephemeral.

Fortunately, the plants in shadier places are behind the early ones, while others are beginning to push up through the drifted mud and will be later yet, so the show will last a while.

According to my notes, bloodroot, observed in full bloom when we went back to the river bank on May 3, were completely finished by Mother's Day, May 8. Other patches, though, were not so far advanced and still in bloom.

Now there is Dutchman's breeches coming up everywhere among the bloodroot—especially at the base of fallen logs—but not yet in bloom. A rather unexpected profusion of nodding trillium is joining in, pushing up through the bare mud.

Figure 34: Nodding trillium

This year I am inspired by Carl Mulden's book *Native Orchids of*

Nova Scotia, and am looking for native orchids everywhere I go. There are 38 species, which is not overwhelming, and I am hoping to find as many as I can this summer.

Last year it was ferns. Next year it might be violets. We like to pursue some quarry besides that which we know we will find. This lends extra anticipation and focus, and drama if we find it.

Best of all, the search demands close attention, eclipsing petty daily worries and concerns. A blessed state of peace and contentment. Try it. It can be orchids or ferns or birds or fossils, or beach glass or anything.

If you don't find what you are looking for, you will always find something, guaranteed. The best discoveries generally come when you are looking for something else.

This river intervale may not be the best place to find orchids, and it's too early anyway. Still I am eagerly looking.

Meanwhile, May 15, the bloodroot is finishing, the Dutchman's breeches are in bloom, and the trillium are coming on strong. The Dutchman's breeches grow in sizeable patches, sometimes among the bloodroot and trilliums, and sometimes in pure stands. They seem to prefer spots at the foot of fallen logs or rotted stumps. Their foliage is light green and ferny; they are delicate-looking and not often taller than 6 or 8 inches. They are closely related to the cultivated bleeding hearts.

Now, to satisfy your curiosity about the peculiar name, the flowers resemble, well, a tiny pair of old fashioned Dutchman's breeches, with creamy white legs and a puckered yellow waistband. They are about one inch long and hang in a line on a single stalk just like pants on a clothesline. They are pollinated by bumblebees, which somehow force their way in through the waistband.

The entire plant yellows and withers within a couple of weeks of blooming, ripening seed before disappearing for the next eleven months. How's that for ephemeral?

By May 29, The riverbank is bursting with life. We see the kingfisher and the pileated woodpecker through the blackflies. Deer tracks meander through remaining patches of mud. By the river

57

bank, the comical handprints of raccoons. And here—Lord—a shredded stump and the big print of a bear.

Bears are on the increase since the farmers began growing corn in the fields alongside the river. We now carry bear spray, though we have never seen a bear except from the car. Anyway now we feel safe and can relax and keep looking for flowers.

Nodding trillium is everywhere—more than we have ever seen here before. It has forced its way up through the mud and is in full bloom. In some places, it almost forms a solid ground cover. Every plant sports a big flower which hangs beneath the leaves so you have to stand on your head to see it. The leaves are big, dark and handsome, proving that mud left by the flooding stream is rich with nutrients.

Finally, the fiddlehead ferns are up, as are meadow rue and the

Figure 35: Fiddlehead ferns (ostrich fern) as trees leaf out

big, rough cow parsnip. There is lots of orange-flowered annual jewelweed. There will be no bare ground this summer.

The trees are leafing out, shade is deepening, and the spring flower show is over a month after it began. In the increasing shade, tolerant species such as jack-in-the-pulpit, blue bead lily, meadow rue and ferns will soon take up the slack.

The hardwood uplands

Rising above the river floodplain, the rich upland hardwood forest shelters ephemerals which, like those of the riverbank, bloom and complete their reproductive cycle before the trees leaf out. Here the native species have escaped the ravages of logging and agriculture due to steep terrain and inaccessibility. Most are the same species found by the river, though some, such as spring beauty, painted trillium and hepatica, prefer the uplands.

On July 4, Mary and I took a drive to Cape Breton to see what we could see, in hopes of finding the painted trillium. Our destination was the foothills of Glendale Mountain.

Years ago, with Mary's cousin Dorothy, we had visited a beautiful mossy glade where a clear, sparkling stream tumbled down the mountain. It was a paradise of ferns and flowers, including painted trillium. This trillium—another Spring ephemeral—is related to the nodding trillium we see by the river. The flower of the painted trillium is colourful, with its three white petals and red centre Also, the flower is on top of the leaves, looking up at you, instead of hiding underneath.

Many things had changed over the years since we were here before. Dorothy was gone, and we had accepted that, but, to our surprise and dismay, so were the woods—recently clearcut as far as the eye could see. The little stream still ran through its green glade of moss and ferns, but we asked ourselves for how long. The tree harvesters had left a surrounding fringe of trees as the law requires, but the little glade would face erosion from the devastation above.

The trilliums were gone.

Sad and disappointed, we turned around and drove along the road back to where the woods were still standing, looking for a place to eat our lunch. We found a comfortable, mossy slope overlooking another nice brook, if you overlooked the washing machines and water heaters thrown into it, and enjoyed our lunch until swarms of black flies drove us back into the car.

I had a quick look around before dashing back to the car. There were plenty of handsome ferns, wild sarsaparilla, and wild lily-of-the-valley, but no trilliums and no orchids to start my list.

After lunch, we continued along and took a logging road that began to climb up the mountain. The road mostly traversed the slope. The woods on the upper side seemed like small second growth, but on the lower side the slope was steep, the ravine was deep, and the trees looked much older.

It looked like this ravine would be where we might find something interesting. It was probably too steep to have ever been thoroughly logged.

We stopped at several places where we looked over the edge, but they were too steep even to go down on foot. What wonders, we thought, must be down in the bottom.

Finally, we found one place where it was possible to descend the bank. I went down alone and promised that, if I found anything interesting, I would summon Mary.

Down in the ravine, the trees were big and old, and there were the normal ferns and plants found under old forest, but nothing out of the ordinary. It wasn't until I was on my way back up to the car that I saw something, on the side of the steep slope.

There it was—the painted trillium— but the flower was brown and withered. We were too late. Still, it was satisfying to

Figure 36: White lady's slipper

have made the discovery. Nearby were several more.

To top it off, looking around, we saw several white lady's slippers and lots of bluebead lily all in bloom.

I called Mary down and we admired the plants until we were

chased back to the car. I can still hear the buzzing of the black flies.

My discovery of the lady's slipper orchid was my first for the season, and I began my list.

May 14, 2023: This Spring, we are again heading into the high-lands, this time up Brown's Mountain in Antigonish County. This is rich hardwood forest, similar to that in Glendale.

There is no evidence of tree harvesting here. In 2005 the province declared a large portion of the Brown's Mountain forest a protected wilderness. Since then the protected area has been ex-panded several times until today it encompasses 7600 hectares of prime upland forest.

We are earlier this year and hoping to find some painted trillium in full bloom this time. We have also heard that we might find spring beauty.

It is a beautiful Spring day, the best so far. The car thermometer says 19 degrees.

The road starts out in good shape, but once we pass the last of the houses on the lower end, it starts to narrow. In places it is stony and rutted. There were a lot of trees that came down along the way from Hurricane Fiona. Many of them must have dropped across the road. Someone, though, bless his heart, has worked hard to cut away the fallen trees enough that a car can get through.

This is the road to Cutie's Hollow, where there is a hunting camp and the start of the trail to the James River Falls. As it climbs, the road passes through mixed woods with conifers, then into rich hardwoods.

By a small spring near the road, a patch of dark green catches our attention. Mary waves me out of the car, so I shut off the engine and get out to have a look. It is something familiar, but it isn't yet in bloom.

It takes us a minute to figure out that it is toothwort, Dentaria. This will be in bloom very soon and is a handsome plant with glossy, three-lobed leaves.

As we are turning to get back in the car, we notice something else just pushing up through the dry leaves. There it is, the painted trillium, and it is just coming up. We are too early this time.

There are quite a few of them. A couple are up enough to have flower buds forming. At least now we've got them cornered and can come back later when they are at their best.

The road continues to climb before beginning a descent to Cutie's Hollow. At its highest point is a steep hillside of ferns, upon which we are delighted to discover a carpet of spring beauty.

Figure 37: Spring beauty

The spring beauty are a delicate white striped with pink, and wave gracefully on thin stalks a couple of inches high. They are growing among last year's ferns, with Dutchman's breeches and a bit of bloodroot.

We were not surprised to find the painted trillium and spring beauty up here in the highlands, but the other two we think of as floodplain plants. We didn't expect to find them here.

As the road descends to the Hollow, the woods alongside get swampier, and hopelessly barricaded with fallen trees. Nevertheless, even though the road is getting very bad, we keep on. We want to see if we can find the trail to the James River falls in case we want to hike it this Spring.

We find it, but it is a shock. No more than fifty or sixty meters from the start, a massive spruce has blown down directly across the trail, with no getting around it. No doubt there will be more all the way. We won't get there this Spring, maybe not this Summer. If someone doesn't do a huge trail clearing job, that trail may never be used again.

We head for home thinking sober thoughts, but stop along the way to look one more time at those spring beauties.

May 24: Back to Brown's Mountain with a friend. Mary is working today and can't come along, but I just have to go back and find those painted trillium again. They should be in peak bloom by now.

We motor slowly up the narrow road, as slow as we can in order to check the banks of the road for interesting vegetation. It's nice for us botany hobbyists to be able to go as slow as we want on a road with no speed limit, and no traffic, and no one behind us forcing us to pick up the pace.

It is another sunny day, and warm for this time of year. The ferns have unfurled since Mary and I were here ten days ago. I am watching for the patch of green toothwort where we saw the painted trillium coming up that day. The flower buds were starting then and should be wide open now.

We find the toothwort patch, no problem, and I bounce out of the car with my camera to get some photos of the trillium.

There are many trillium up all right, but something is wrong. There are no flowers.

And then I see. The flowers are hanging beneath the leaves. These aren't painted trillium at all. They are the nodding trillium that we have already seen all over the place on the riverbank.

As I look around, I see the Dutchman's breeches and even bloodroot all around. Nice, but the very same stuff that we see on the riverbank close to home.

On a day like this, I can't complain if all I find is nodding trillium, but I feel foolish and disappointed. When I first saw the trilliums here, I forgot that when the nodding trillium is first forming its flower buds, as they were that day, the flower is on top of the leaves. That is why I thought they were the painted. It is as they become more mature that the flower nods.

We continued up the road to see if the spring beauties were still blooming. They were—just as nicely as they were ten days ago. There seemed to be much more Dutchman's breeches mixed in with them this time, and more ferns.

We climbed up to the hilltop above the road and the spring beauty was still everywhere up there. The ground seemed drier than down below, however, and there were no nodding trillium.

It occurred to me that when I had seen painted trillium in Cape Breton, it had been in drier locations. I thought that if it was found in these hills, it might be farther up the road, where it was drier yet.

We drove along higher and higher up the mountain to where there was a little pull-off and a sign announcing that everything past this point was a protected wilderness area. The terrain did look drier here, and a wilderness seemed a likely place to have a look around. I really wasn't hopeful, but went looking.

Then, there it was—the painted trillium. And there was another and another. Their graceful white flowers with the crimson centres were full out and looking up at me just like they were supposed to.

Beneath each flower were three sleek, wavy, purplish leaves that set *Figure 38: Painted trillium* the flower off perfectly.

The leaves of the painted trillium are so distinctive that I will never again confuse them with the nodding.

My friend and I took photographs and we returned home highly pleased. It will be our last time out this year looking for ephemer-als.

With the trees in full leaf, by the first week of June, the show is over. We will anticipate it eagerly next spring.

It was these first happy flowers of spring, the bloodroot and its companions blooming on our little bit of river bank, that first cap-tivated Mary and me and got us started on wildflowers. From these fragments, these scattered remnants of what was once the richest forest in the province, one may imagine the glory that existed be-fore colonization.

Over one hundred and twenty years ago, in 1901, C.B. Robinson of Pictou Academy wrote describing the rich ephemeral flora found on the stream and river intervales of Eastern Nova Scotia, and wondered why it wasn't taught in the schools. In the same breath, he lamented the increasing loss of wildflower habitat due to logging and agriculture.

Sad to say, even then the writing was on the wall. Since Robin-son's day, logging and clearing have been mechanized to more effi-ciently eat up wildflower habitat.

On the bright side, clearing on intervale land may have gone just about as far as it can go. There are laws that prohibit clearing land too close to waterways. River bank acreage may even be increasing as agriculture languishes in this age of artificial intelligence.

Today, the importance of protecting the sensitive flora of the hardwood highlands is accepted, and large tracts are beginning to be set aside as reserves and wilderness areas. So, the future is hopeful; but for now we must be satisfied with a glimpse, as if looking through the wrong end of a telescope, of the wonders that must have been.

Figure 39: Clintonia - a companion to all

3: Mayflowers

Figure 40: Mayflowers

Now, what about the mayflower, our provincial flower, and maybe the sweetest smelling of them all. We don't want to miss the mayflower. Around here they do indeed bloom about the middle of May.

The mayflower is a very unassuming plant, carrying on almost unnoticed in some of the poorest of soil. It is evergreen, but lies flat to the ground, half hidden by leaves and pine needles, until in May its fragrant pink and white flowers peek out at the sun.

Our fondness for this most shy and retiring of wildflowers is a tribute to our people. As early as 1820 the mayflower was seen to represent Nova Scotia society, as a symbol of humility and high achievement in the face of adversity. This representation was embraced province-wide. The mayflower motif was embossed on the Lieutenant Governor's chain of state and on tunic buttons of the militia. Its image appeared on coins and postage stamps, while citizens wore mayflower pins and flew mayflower banners. At least two newspapers were named for the mayflower.

Figure 41: The mayflower on an 1856 halfpenny token

In 1901, in an act of the Provincial Legislature, the mayflower was declared to be the floral emblem of Nova Scotia, and to have been so since time immemorial.

Incidentally, the mayflower is also the state flower of Massachusetts, but they didn't claim it until 1918. We got it first.

Our floral emblem is an ericaceous plant, meaning that it is in the blueberry family along with the blueberry, bearberry, huckleberry, rhodora, lambkill, the wintergreens and many others.

The ericaceous plants, unlike most others, prefer an acid soil. They have certainly come to the right place in Nova Scotia, where they thrive from one end of the province to the other in the sour soil of mossy bogs and barrens and scrawny woods and clearings. Ericaceous plants tend to be found together.

Mayflowers are partial to gravelly or sandy soil, and moss. In moist ground they can tolerate considerable sun. In drier places they like some shade. We find them along the edges of old woods roads and trails. They almost always grow in the moss among the leatherleaf, the labrador tea, the blueberry, the wintergreen, and others of that sort.

Figure 42: A patch of mayflower

The leaves are distinctive, and once you learn to recognize them you see them more and more. The flowers are bashful and shy and often we see the leaves before noticing the flowers.

In Antigonish County, where we live, the soil runs to shale and clay and we don't know of many spots where the mayflower is happy. There is one place, though, behind the dunes at Pomquet Beach Provincial Park, that is outstanding.

Mary and I visited as many provincial parks as we could last summer, and hiked their trails. This is a very good way to look for native plants, although by mid-summer they are looking a bit used. Pomquet Park being almost in our backyard, we were able to visit it early when the mayflowers were blooming at their best. We re-turned several times in Summer and Fall, and though the mayflowers were done, it was lovely every time.

Mayflowers are only one of the wonders found there. The park people have constructed beautiful boardwalks through what they call the Pomquet Dune Slacks—the forest inland from the seashore and the dunes. This is a somewhat stunted forest growing in what is essentially beach sand covered in moss and carpets of bearberry, bunchberry, blue bead lily, and all the ericaceous shrubs including mayflower.

Figure 43: Bearberry

Later in the summer, we found numerous specimens of the pink and the white lady's slipper. The trees are predominantly red oak, with red maple and birch. Oak forest is almost nonexistent in Antigonish County, and on the county soil map, this tiny piece of forest has its very own designation—Po.

Fall colour is spectacular in Po. Of all the trails we did this summer I think this is the best, and we can drive there in twenty minutes. How good is that?

Our first visit to this place to see the mayflowers was May 16.

The mayflowers were in the peak of bloom, but that was not all. Low mats of bearberry glistened in the sun with clusters of flowers —tiny pinkish bells—rising above the leaves.

The bearberry is another ericaceous plant and grows prolifically in the sandy soils at the other end of the province, but is almost unheard of here. It has small, glossy evergreen leaves, and after flowering is sprinkled with bright red berries—hence the name.

Another name for this plant is kinnikinnik, derived from an Algonquin word. It is said that the native people used it in smoking mixtures for their pipes.

We are not finished yet. Besides the mayflower and the bearberry, there was the pyrola and the prince's pine, and something that looked like Solomon's seal but turned out, once it bloomed, to be the starry false Solomon's seal.

On that day in early May, lots of things were coming up, but too small to identify, and the huckleberry was not yet in bloom. We would have to come back again.

We did go back again—several times. The first was June 21. It was summer solstice day. There were ominous looking black clouds in the distance but it was sunny in Pomquet.

Figure 44: Pink lady's slipper

By now, the huckleberries were covered with red flowers. The mayflowers were finished blooming and were now almost hidden beneath bunchberry, wild lily-of-the-valley, and wintergreen; but, looking closely, we could see healthy new growth.

71

Scattered throughout the woods were lots of pink lady's slipper —most just past their prime, and a lot more poison ivy than we ever knew was there. I suppose that is because the poison ivy leafs out late, after the mayflower blooming season is over, and we had never noticed it.

We stayed out of it, and when we returned home we found that the big black clouds we had seen from Pomquet had burst over our

Figure 45: Bunchberry

place, and everything was soaked.

The last visit we made to Pomquet was on October 27 when we wandered through the incredible Fall colours. Quite a few trees were down from Hurricane Fiona, but the park staff had cleared them off the trails promptly, I guess while we slept.

Mats of juniper and bayberry were brilliant green. The huckleberry bushes now bore tasty, shiny black berries. Everything was glowing yellow, red, and orange—including the poison ivy.

Visiting a favourite site repeatedly, as we did at the Pomquet park, teaches that all of nature is a moving picture. Each time we visit it stops to give us a look, but when we leave and our backs are turned, it starts up again. The next time we visit, things have moved on and everything has changed. The seasonal cycles of life in the plant world are fascinating, ongoing and eternal. Nice places are worth visiting more than once.

Besides Pomquet, we visited another place where we had seen mayflowers in the past. This was on a drive to St. Peters in Cape

Figure 46: Starry false Solomon's seal

Breton, where we wanted to visit a garden centre. It was May 1st, and cold. Of course the trees were all leafless and dead-looking, and it seemed unlikely that anything could be alive and blooming on such a day. Nevertheless, as we were passing, we swung in to have a look.

This was just about the least likely place to find wildflowers that you could imagine—back of the buildings at the Strait Regional Hospital in Evanston. Behind the main buildings and the parking lot, in between sheds and service buildings, the ground was mossy, wet and soft. The groundskeepers kept the grass short, barely higher than the moss.

Figure 47: Blueberries with poison ivy

Flat on the ground in the grass and moss were patches of miniature mayflowers—just two tiny leaves on each plant—every plant in bloom. How could that be? Even a dandelion couldn't survive in that ground.

Figure 48: Darker pink mayflower

Farther back, at the edge of mossy woods, there were more mayflowers. Here, the soil was better and they got some shade. They looked happy and grew to normal size. But they weren't blooming.

74

A few had flower buds but it appeared that flowering was not high on their agenda. This is easily explained.

If you have done much work with plants, you will have discovered that plants don't bloom to be beautiful. They bloom to reproduce. The more stressed they are, and the more worried about their own survival, the better they bloom. Those little mayflowers in that wretched ground, in full sun, probably didn't expect to be around next year. They weren't going to pass up a chance to bloom and set some seed. The luckier plants at the edge of the woods had no such worries and were very lazy bloomers.

Blooming and ripening seed takes energy and comes at the expense of vegetative growth. A young, vigorous plant growing under good conditions often chooses to grow bigger rather than bloom.

Mary and I both worked at a garden centre for most of our careers and frequently observed this phenomenon. In fact, it works to the benefit of garden centres. Most of the trees, shrubs and perennials for sale each Spring were disturbed and stressed in one way or another. Often they had been dug up or repotted or were crowded into pots a little smaller than they would have liked. They responded with abundant bloom, which made them easy to sell.

Very often, people who had purchased these immature plants came back a year or two later to report that the plant was growing nicely but it never bloomed. Was there something wrong with it or something they should be doing? We had to go through the whole explanation and ask them to just be patient.

What did worry us was when a customer came in to boast that his small fruit tree, planted a year or two before, was covered in flowers. He figured this meant a lot of apples, but to us, it meant that, for some reason, the tree thought it was dying. It often was. It shouldn't have been blooming that young.

We finished our drive to Cape Breton with a tour of Isle Madame, where we hadn't been for many years. By then, the weather had turned even greyer, wetter, and colder, and we elected to keep driving rather than to get out looking for mayflowers. We got lost on the roads and saw a lot of Isle Madame that we didn't know existed.

We were amazed at how self-contained the place was, with stores, restaurants, houses crowded together in the French style, and an enormous fishery. Mary spotted ospreys in nests on top of telephone poles—a bird we don't see very often anymore around Antigonish.

There were probably mayflowers all over the place, but we couldn't see them from the car. We did notice, through the fogged up windows, lush mats of low seashore shrubbery such as bayberry, winterberry and roses, and there was undoubtedly much more.

Isle Madame isn't far from where we live, and we'll put it on our list for next year.

By the way, the mayflower is one of those wildflowers that are thoroughly wild and doesn't take kindly to cultivation. Let's leave it wild and not try to dig it up. Where it is plentiful it may be impossible to forgo picking a spray of fragrant flowers to put in a jar, but snip off sprigs.

Don't yank up the whole vine out of the moss. That would be a horrible thing to do to our provincial symbol of humility and high achievement in the face of adversity.

4: Gypsum cliffs

Figure 49: Gypsum cliffs

In June, as the shade deepens in the forest and we wait for the ferns to grow, is a good time to go exploring in the gypsum. The gypsum is a world of its own, and rarely visited, but for the eager botanist.

No botanist could be more eager than were Merritt Fernald and his crew in the 1920s. Fernald writes:

> Pease and Long, having spent the preceding day in a hope-less barren, had earned the novel assignment for the day, the calcareous valley of 5-Mile River with its great, fantastic

white cliffs of gypsum. To be sure, they had to get up by 5 o'clock and their return train would not get them back until after dark and long after supper-time. But what of that!

When, toward 9 o'clock, the 5-Mile River party came in, they were a tired, hungry and rain-soaked pair. They had been out since early morning in the richest spot of the summer and their sneakers and clothes plainly showed the result of a day of enthusiastic exploration of the knife-sharp pinnacles and unyielding talus and crests of gypsum.

Nova Scotia has the most extensive gypsum deposits in the Northeast. To the detriment of wild flora, gypsum has been mined here for over 200 years for the production of plaster and drywall. A Government of Nova Scotia bulletin claims that the province is the world's most productive gypsum mining area. I wish we were number one in something else.

Nevertheless, there are many places in the province where large gypsum deposits remain.

Chemically, gypsum is calcium sulphate deposited by receding oceans in prehistoric times. It is soluble, and may lie invisibly beneath the soil, where the only hint of its presence is conical sink holes where it has been dissolved by moving water.

It is most evident, though, where it erupts above the ground in the stark white cliffs and spires called karst. Most of us have seen karst in our travels. Probably the most familiar example is the white cliffs alongside the St. Croix River on the way to Windsor.

The karst outcroppings are heavily sculpted by water and eroded by sun and wind. They form a tortured landscape of sink holes and narrow ridges, supporting a tangled forest of runty trees with no commercial value. This may be the only true wilderness left in Nova Scotia, and a refuge for endangered species. It's a rare human being, other than the botanist, who ever ventures in.

John Erskine expresses it this way:

The white weathered gypsum on the surface is now of little value and remains untouched. At its edge, too, farming stops, for the surface is sculptured into a gigantic honey-comb of vertical caves through which rainwater seeps down towards sea-level, and in the roughest areas even cattle cannot graze, for the vegetation is scanty and the danger great. These few small ears remain almost unchanged by man, and in them one glimpses a native vegetation which has been driven wholly from the richer lands.

Where gypsum occurs, above ground or below, it has a profound influence upon soil and growing conditions. Decomposing and dissolving gypsum releases abundant calcium and sulphur to the soil. Either or both of these can be deficient otherwise.

The high concentration of calcium ions protects the soil from acidification by acid rain, and permits the growth of a unique class of plants termed calciphiles. Two are the rare yellow lady-slipper orchid, and the very rare ram's head lady-slipper orchid.

The bulblet bladder fern, which elsewhere is rare, is common in the gypsum; and if the maidenhair fern is ever seen in Nova Scotia again, it will no doubt be there, too.

Venturing into a sizeable area of gypsum karst is a good way to get lost, so take a compass. Weaving around for hours between sink holes, looking for the yellow ladies-slipper, will do it. And have some respect for those sink holes. The sides are steep and crumbly and sometimes there is a pool of azure blue water in the bottom. If you slide into one of those you may never climb out.

Fortunately, along the ridges between the sink holes there are scrawny trees and shrubs just the right size for hanging on to.

Many species of grasses, ferns and flowering plants are found almost exclusively in the gypsum. It seems that a high percentage of remaining endangered black ash—the ash the Mi'kmaq make baskets from—are there, too.

Between June and September 2014, a plant inventory survey was carried out at 36 gypsum locations across the province by David Mazerolle, Sean Blaney, and Alain Beliveau of the Atlantic Canada Conservation Data Centre in Sackville, New Brunswick. Their study uncovered 857 species, including 83 that were provincially rare. Of these, 42 were vulnerable, 28 were imperiled, and 13 were critically imperiled. Of the 83 rare species discovered, 30 were calciphiles, known only from the gypsum.

The three researchers doing this study also made maps, took photographs, and added interesting comments which make me want to get right out there this Summer. I believe I'll try it first around middle of June, because that is when I have seen the yellow ladies-slipper blooming in the past.

John Erskine tells us:

> It is in early spring that the flowers of the gypsum are at their best, for the humus is thin and the plaster rock holds up little water. Before the gypsum "chimneys" become parched, some early flowers live out a few brilliant weeks.

I guess you could say the middle of June is "early spring" in Nova Scotia.

Figure 50: Spring in the gypsum

The yellow lady's slipper orchid is what lures us into the gypsum. I discovered them years ago when collaborating with Henri Steeghs on his popular video "Treasures of the Old Forest", which, incidentally, is available for watching on YouTube, posted by David Patriquin. Though the yellow lady's slipper is by no means the only remarkable thing in the gypsum, we hoped they were still there.

Not wanting to miss them, and a little unsure of when they would be in bloom, I jumped the gun and went alone into the gypsum on May 19.

This particular escarpment rises prominently alongside the South River, near where we live. When you are beside the white cliffs, they are much steeper and taller than they look from the highway. There is no easy way in.

I was pretty sure that Henri and I had climbed up the east side when we were filming for the video, but the west end looked more accessible, and only required climbing over a couple of fallen trees. I figured that if there were orchids at one end, there should be orchids at the other.

Once over the trees I was into the gypsum, which meant some climbing at first. The gypsum at that time of year was damp and crumbly, with a maze of ridges and sinkholes. It took some care to not misstep and skid down the side of a ridge, possibly erasing some rare wildflowers on the way.

Inside the gypsum, it is slow going but not difficult. You just have to look ahead as you go to plan your route so you don't end up at a dead end.

It is always peaceful picking your way slowly along while scanning the sides of the sinkholes for ferns and orchids. Pick out which little trees you are going to hang on to as you go along.

In a big gypsum formation, it is easy to get lost from all the twisting and turning, but in this case I could hear the faraway sound of traffic on the highway, so knew where I was.

I discovered that I was too early for the lady's slippers, or there weren't any. It was very early. The trees and dogwood shrubs and hobblebush were budding, but not leafing out yet.

Sunny faces of gypsum were studded with small dark clumps of some kind of plant that was just coming to life. There was no sign of orchids—neither leaves nor flowers—and no ferns yet. On the ridges there were clumps of the wild sarsaparilla just erupting into growth, but not much else. Still, it was sunny in there and the birds were singing.

I wandered around until I was tired then made my way out.

As I dropped from the last gypsum hillock into the flat hayfield alongside, I noticed that not far away, at the edge of the field, there appeared to be a deep canyon. I went to have a look.

It was what it seemed, and very steep, with gypsum cliffs rising on every side. At the bottom were big hardwood trees of the same species, but much taller and straighter than the scrubby specimens up in the hills.

It was tantalizing to consider going down for a closer look, but it seemed that it would be tricky sliding down the side of the canyon and climbing back out again. And I was tired.

There was something, though—some kind of plant forming a sizeable patch of brilliant green—way at the bottom of the cliff. Maybe it was a patch of some kind of rare orchid or fern. If I was any kind of botanist I should climb down and have a look.

I spent a while resting and casing the situation, and noticed that if I walked along the rim of the canyon for a ways, it became much less steep and easier to get into. So that is what I did.

The trees at the bottom of the canyon were well spaced and easy to walk through, and along the way I spotted almost every kind of fern. Spring was much more advanced down here. There was star-flower and wild lily-of-the-valley and Canada honeysuckle in bloom.

When I reached the patch of bright green, I was puzzled. The plants were lush and six or eight inches high, with stalks of quite large flowers about to open. It certainly wasn't orchids. In fact it was nothing that I had ever seen.

My first thought was that it was an ornamental plant that had escaped from someone's garden into the woods—something that happens all too frequently. Still, it wasn't anything I was familiar with.

The leaves were decidedly three-parted and my second thought was that it might be poison ivy and I was afraid to touch it. But no, the leaves were all wrong and poison ivy didn't have flowers like that.

Since there was a lot of it, I did what we do in cases like this. I took a plastic bread bag out of my pack and stuffed in a small clump with leaves and flowers, to identify at home.

At home, I put the clump in a jar of water on the windowsill, waiting for the flowers to open. If you can find flowers, you can identify a plant using the key in the Flora.

That is what I did, but when I keyed it out, it came out to toothwort. I know toothwort and it is a handsome plant. I should have mentioned it among the Spring ephemerals. But this wasn't toothwort.

The next step was to take down the Petersen Guide to Wildflowers and start looking at pictures.

Finally I found it—large toothwort! No mention of large toothwort in Roland's Flora, so I went for the latest two-volume set

Figure 51: Large toothwort

by Marian Zinck. Large toothwort was in there, but from only one location in Nova Scotia.

That was a nice feather in my cap, I thought, to find something so rare.

When I boasted of my find to Sean Blaney, a career botanist in Sackville N.B., though, he bent the feather a bit, informing me that it is now found in various parts of the province.

Hoping that there were yellow lady's slippers where I was searching in the gypsum, but that I had just been too early the first time, I went back three weeks later, June 8, with Mary. The dog-wood shrubs and hobblebush, and all the trees, were leafed out in the pits and ridges, the wild sarsaparilla was tall and blooming, and plants were growing all around.

Figure 52: Yellow lady's slipper with fleabane

The dead-looking little clumps that I had noticed scattered all over the cliff face on my first visit had come to life and were blooming. Now I could examine their flowers and identify them. Some were hyssop-leaved fleabane, a small daisy-like flower. Others were the little balsam ragwort with yellow flowers. Both, when found, are almost always in the gypsum.

We wandered slowly, which is all you can do in the gypsum. The vegetation was very lush. It seemed everything was growing there except the lady's slipper.

We had almost given up when there it was, hiding in a sinkhole: two of them.

We took pictures and carefully made our way down for a closer look. One was good sized, but the other very small.

The yellow lady's slipper is small anyway—not nearly as big as it appears in pictures. A good-sized one is no more than eight or ten inches high, with a flower about the size of a loonie. It is not nearly as big as the pink lady's slipper.

Anyway, I could tick another orchid off my list, and we resumed our search, expecting that we would find more. That day, though, we didn't, and I hoped it wasn't because they were dying out in this location.

The other end of the gypsum escarpment—the east end—appeared to be the hard one to climb. I was pretty sure, though, that I had seen lady's slippers there when Henri and I were making the video. I went up this way on my next visit a few days later.

The way up proved not so bad. It was steep, but there was an animal trail and small trees to hang on to.

I wasn't very far up when I spotted the first lady's slipper beside the trail. From there on to the top, lady's slippers were scattered here and there, sometimes singly and sometimes in groups of two or three or more. There was a flower on every one. They seemed to choose spots a bit down on the sides of sink holes where they had a bit of shade and a pocket of rich organic soil.

The cliffs were much taller here than they had been at the other end, and the view quite breathtaking. To the south were farms and fields and the old highway to Antigonish. To the north was an impossible labyrinth of peaks and pinnacles of gypsum, clothed sparsely with maple, birch, spruce and pine, descending to cliffs dropping into South Side harbour.

Further wandering up on top resulted in the discovery of even more lady's slipper. It was as if not a single human being had set foot up here since we were up twenty years before.

That could very well have been true. There was no reason why the ordinary person would go to the trouble to climb up. It was slippery, there was no timber, it was too rough for hunting, there was no fishing, and who cared about a few orchids?

Well, we did and were glad no one ever came up to trample them. They even seemed to be increasing.

Now that I knew where they were, I went up twice more—once with Mary, on June 10, and once with our daughter Margaret on June 15. Both times, the lady's slippers were still blooming beautifully. They stay in bloom a long time.

We were getting a bit jaded by yellow lady slipper at this point, and turned our attention to everything else that was happening on top of the gypsum, such as luxuriant mats of blue-green juniper, as well as prostrate, wind-blasted spruces, pines, and firs.

Figure 53: Twinflower in the gypsum

Bracken fern had begun to fill in empty spots and Margaret found a large mat of twinflower up high and dry on a gypsum hump in full sun. This creeping plant usually grows in the moss of shady forests, but was nevertheless blooming profusely in the gypsum, looking a little sunburnt but with much darker flowers than usual. As the name implies this plant blooms with pairs of tiny pink bells on stalks. Of the thousands of plants Linnaeus named for others in the 1700s, this is the only one he named for himself—*Linnaea borealis*.

On the way out of the cliffs we passed perfect clumps of blue-eyed grass in the moss and gravel near where we had parked the car. Blue-eyed grass, a small iris relative, is as choice a native wildflower as they come, and you don't have to do any climbing to see it. Like the purple violet, it is too often written off as common.

Early spring isn't the only good time to visit the gypsum hills, by any means. Maybe the exposed surfaces become hot and dry, but there is plenty of shade and moisture on the

Figure 54: Blue-eyed grass

steep slopes of the sinkholes. The ferns can be spectacular, as well as shade-loving herbaceous species and uncommon shrubs like the hobblebush.

It is cool down in those sinkholes in the heat of summer. There is one not too far from here that people call the ice cave. Cold winter air is stuck in the bottom all summer and can't get out. There are chunks of dirty old snow year round. It's a good spot to remember during a heat wave. You will feel the cold air by the time you are halfway down and the dirty snow will keep your beer cold.

Patches of forest adjoining the gypsum are good places to explore and may surprise you with another orchid or a Pyrola.

I am worried about the gypsum. Gypsum mining is big business in Nova Scotia, and large open-pit mines are at work in various parts of the province. These mining operations are interested mainly in hard subterranean gypsum deposits, and not so much in the cliffs and spires of crumbly gypsum karst where the wild-flowers grow. They bulldoze these to get them out of the way.

The 2014 gypsum survey by the botanists at the Atlantic Canada Conservation Data Centre discovered numerous rare and en-dangered plant species in the karst. Many of the sites surveyed had never been systematically explored before. Their discoveries are proof that the rugged and nearly inaccessible outcroppings of karst in Nova Scotia are a haven for threatened species.

The frightening part is that most of this karst topography is un-protected and located on private land. Large mining companies own the rights to almost all undeveloped gypsum deposits in the province, and have no use for karst or yellow lady's slipper. The gravity of the situation is expressed in the survey:

> Large expanses of natural gypsum karst have been lost to other land uses, predominantly quarrying, farming and forestry. Gypsum mining has a long history in the province of Nova Scotia and large open pit mines have already re-moved many of the most significant examples of gypsum landscapes, with mining companies presently owning large portions of undeveloped gypsum land.

The survey also points out that at present only one percent of sensitive gypsum outcrops are located on protected land. The au-thors urge the protection of all remaining areas of gypsum karst in the province, and pinpoint several locations of top importance, among which are several near where we live.

Breaking news—Spring 2023. The government is listening. One of the sites deemed of top importance in the gypsum survey, a kilo-metre or so of old growth forest that somehow was never cut, has been protected. This floodplain forest lies along the Southiver near our home in Antigonish county, and shelters an intact understory of forest flowers.

It was always a mystery to us how it survived year after year, while the rest of the land along the river had long ago been logged and turned into farmland. We were certain that it was only a matter of time.

This spring, driving to the spot to have a look at the bloodroot that was beginning to bloom, we noticed yellow signs nailed at intervals to the trees. Finally we became curious enough to read one.

Our hearts leapt with joy. The signs proclaimed the area a nature reserve. There was to be no tree cutting, no motorized vehicles, no camping, no open fires or smoking, no hunting or trapping, and no littering.

Later that day, in the official list of Nature Reserves, I found this:

> South River Nature Reserve protects four small parcels along South River, where the river meanders through the biologically rich lowlands of southern Antigonish County.
>
> The reserve includes a mix of rich river floodplain and gentle forested slopes. It provides habitat for wood turtle, a listed species-at-risk. At least five at-risk plants occur here, including black ash.
>
> The reserve protects the only provincial lands on South River in a landscape with much land use and disturbance of natural habitats. It is bounded by Dunmore Road along the western side and South River along the eastern side.

Figure 55: Our turtle

And just to inaugurate our new Nature Reserve, Mary and I found a wood turtle that day.

Figure 56: Cinnamon fern

5: Wetlands

The Bog

Figure 57: The bog

At the opposite extreme to the sun-baked gypsum karst is the bog —another sort of wilderness. No one goes into the bog either except the berry picker or the botanist.

When Merritt Fernald, Lily Perry, Charles Weatherby and others were coming to Nova Scotia from Harvard University a hundred years ago, they weren't interested in our forests. They had bigger ones in Massachusetts. They weren't interested in our wildflowers. They had better ones in Massachusetts. They were interested in our lakeshores and bogs, where they might find the disjuncts they craved. We have better bogs than they do, and for this we are lucky.

Fifteen percent of the land area of Nova Scotia is comprised of wetlands—peat bogs, fens, marshes, and swamps—with most of it peat bogs. Worldwide, peatlands make up 3-4% of the earth's surface. Large areas of bog around the world have been lost to flooding, draining, and mining for fuel. We can't afford to lose any more.

Sphagnum, the common and hard-working moss of the peat bog, plays an ecological role on this planet more important than all the trees put together. Trees, as we know, take up and store carbon, doing their part to mitigate climate change. Sphagnum takes up more.

Edward Struzik, in his book *Swamplands*, adroitly subtitled "The Improbable World of Peat", puts it in perspective:

> The Amazon and other rain forests get well-deserved attention for the amount of carbon they store and for the exotic plants that are harvested and sometimes pillaged for their medicinal properties. But peatlands sequester 0.37 giga-tonnes of carbon dioxide (CO_2) a year—storing more carbon than all other vegetation types in the world combined.

If that doesn't cause your jaw to drop, that is just the start. Sphagnum absorbs and stores water, mopping up floods when they happen, and releases water again during times of drought. Water filtered through peat is cleaned of toxins and pathogens and purified. The vast bogs of the North and the Arctic tundra are the breeding grounds of millions of birds, and home to plant communities found nowhere else. Small bog-dwelling mammals are at the bottom of a food chain leading to wolves and bears, while musk-ox and caribou roam and feed on the tundra.

Of these things, of the immense value of bogs and wetlands to mankind and the natural world, we have been shamefully blind.

A northern bog, such as we have here, usually forms in a glacial depression in the bedrock with no inflow or outflow of water. The fact that nature has given us sphagnum moss is a miracle. Edward Struzik puts it this way:

> Most of the peatlands in the northern hemisphere and far southern regions are dominated by mosses such as sphagnum, which Canadian ecohydrologist Mike Waddington calls a 'supermoss'. It's an exquisite looking sponge that can hold between 15 and 25 percent of its weight in moisture. When it grows and spreads out as a mat over water, it can support the weight of a moose, a bear, or a small forest.

A sphagnum bog is a layer of green moss growing on top of a much deeper layer of wet, compressed dead sphagnum, or peat. The bog has no inflow or outflow of running water; it is wet only from snow and rain, which it soaks up and stores almost 100%, creating what is essentially a solid puddle of perpetual water. This puddle won't burn. If the surrounding forest catches fire, it stops here.

In wet weather the moss sops up water and reduces or prevents flooding. In dry weather, it slowly lets it go to the trees so they don't catch fire in the first place. It took us this long to catch on.

Until very recently, bogs of peat have been thoughtlessly squandered and exploited worldwide. Considered of no value to mankind as they are, they have been drained to make way for agriculture and the building of cities; excavated for fuel, animal bedding, and garden supplements; and flooded by vast hydroelectric projects, along with hundreds of thousands of hectares of rivers and woods also of no value if you don't count ancestral hunting and fishing grounds, or wildlife.

As early as the 1500s, in populous European countries whose ancestral forests had long since vanished into the fires, blocks of peat, dug from the bogs, were dried and burned for heat. This was poor man's fuel. The more affluent burned coal and whale oil.

In England, Ireland, and Scotland, and particularly the Netherlands, peat deposits could be thirty feet deep and have been mined for centuries. In many countries, peat is mined to this day.

Peat has fuelled furnaces, trains, and power generation; been used for insulation, bandages, diapers and tampons; smelted as a source of iron; used in the manufacture of explosives and for horse bedding during the world wars, and mined as a garden amendment. In Canada, peat was tried as fuel for trains, and is bagged and sold continent-wide for use in the garden.

Large areas have been drained for agriculture and building projects, and larger, almost unimaginably larger, areas flooded for hydroelectric developments such as James Bay in Quebec and, more recently, Muskrat Falls in Newfoundland. Despite all this, thanks to our huge land mass and small population, a good percentage of bog land in Canada survives.

World-wide, this is not always the case. Sphagnum bogs once appeared to be inexhaustible and more valuable dead than alive. Suddenly, both times and the climate are changing.

With climate change no longer debatable, and the frightening increase of fires, floods and drought, we want our bogs back. Sadly, the chickens have come home to roost. Most wetlands cleared over the last couple of centuries for trees or crops, or filled to build houses and cities, will never be seen again. Floods may continue to pour through city streets and fires burn through forests that were once wet bogs. Peat itself, where it has been drained and dried, is a dangerous source of fires, and can ignite and burn underground for months or years.

We are belatedly coming to grips with all this, and worldwide the tide is beginning to turn. Money and resources are going into desperate programs to re-wet dead and moribund wetlands and bogs, and turn them green again.

From above, a bog resembles a soft green lawn or golf course, but, zoom in and you will find a complex living thing. Unique populations of birds, animals, insects and butterflies are found in bogs. Many plants are found only in bogs, including bizarre, insect-eating, carnivorous pitcher plants, sundews and bladderworts. Several of the choicest native orchids in Nova Scotia are found in the bogs, as are certain grasses, rushes and sedges, like the curious cotton grass.

Mary and I have a bog. It lies just off the trans-Canada trail near our cabin in Cape Breton. We found it one summer because we spotted cotton grass and cranberries.

Figure 58: Cotton grass

Cotton grass looks just like it sounds—a spiky, grass-like plant with bits of cotton fluff caught at the tips of the spikes. If you find cotton grass, you've got yourself a good bog. We had to wade in to see the cranberries, then spotted the bog laurel, the Labrador tea, and the huckleberry, low shrubby members of the same family.

We saw no orchids, which was disappointing, and no pitcher plant, which was surprising, but we returned in fall to pick cranberries and made a note to visit earlier next Spring.

In Spring of 2022, the earliest visit we made to the bog was June

25, which is still Spring in the bog, and things were just waking up. There had been a lot of rain the day before and the moss was very wet and floating on water. Mary wasn't wearing the right boots and waited for me up on the trail while I stepped in with my rubber boots.

It was very soft—just like wading in water—and it felt like I might go in over my boot tops. After a few steps I found I was only in a little over my ankles and quit worrying.

Exploring in a bog is a unique experience and

Figure 59: Bog laurel

you learn to do it very slowly. You must watch where you step so as not to go in too deeply. Try to step on the humps and avoid the dark spots. Watch out for delicate plants such as sundews and orchids so as not to step on them. There is nothing to hold on to in the bog, and if you lose your balance you're in for a dunking. Walking in a wet bog leaves deep footprints and you wonder if you are doing serious damage to the moss. The next time you visit, though, the moss has sprung back up as if you were never there.

I didn't see orchids or sundews, only some suspicious or-chid-like leaves, but shrubs were blooming. In small hummocks here and there in the moss, low bushes somehow manage to grow.

Virtually everything that grows in the bog depends on mycor-rhizae. Mycorrhizae are thin, almost-invisible fungus threads that travel far and wide beneath the sphagnum moss.

The bog is a very sterile place. The only nutrients that enter the bog come from rain and snow, which supply next to nothing. There are not enough nutrients in any given place to support plant growth, but mycorrhizae cooperate to transport nutrients from a large area into plant roots, in exchange for products of photosynthesis which a fungus cannot make. It's a very tidy arrangement which seems to suit especially the members of the blueberry family, and orchids.

Figure 60: Labrador tea in bloom

Shrubs in the bog are found in humps occurring here and there over the moss. Here their roots get enough air, water is not a problem, and mycorrhizae set the table. Today, Labrador tea and huckleberry are blooming, bog laurel not quite, and no orchids are in evidence. Cranberry is blooming everywhere, promising good picking in the Fall.

Figure 61: Pink lady's slipper

On my way climbing out of the bog I discovered one small pink lady's slipper in bloom and met up with Mary to gloat. She wasn't impressed. While I was in the bog she had taken a look on the opposite side of the trail in mossy woods, and found dozens of pink lady's slipper in bloom—big ones.

Well, it wasn't really a competition, and I was excited to see them.

Our next trip to the cabin and the bog was July 5. This time, the weather was beautiful and sunny and conditions were drier. Mary wore her rubber boots and I wore mine.

We both waded into the bog and immediately spotted the orch-ids. We were right on time. They had appeared out of nowhere since our last visit, and stood above the moss with bright pink flowers on thin stalks. Some plants only had one flower, some had several. They stood eight or ten inches high and there were maybe several dozen.

Consulting Carl Munden's field guide, we identified them as the

Figure 62: Grass pink orchids

grass pink orchid and were thrilled to have discovered another orchid—especially such a delicate and pretty one. We carefully ap-proached as close as we dared to have a good look and take pic-tures; but again, the bog was so soft we were afraid that we were doing more harm than good. We more or less tiptoed back onto the trail.

On the way we did notice that small clumps of bog laurel were in bloom and that many of the cranberries were already sporting small green fruit. I also noticed that the orchid-like leaves I had seen on earlier visits were larger and still orchid-like, but with no obvious flower stalks coming on, and no trace of spent ones. Apparently if they were going to bloom they hadn't done it yet. There are orchids other than grass pinks that grow in bogs and I still hoped it would be one of those.

We were back to the bog for the last time on August 15, mostly to see how the cranberries were coming along. We could see there was a lot more happening, though, and the bog was a little drier in August than it had been, so we went in to have one last good look around.

Figure 63: False holly

The grass pinks were done for the season and gone without a trace. Those mysterious orchid-like leaves we had seen before were still plentiful. Many of them now were large and turning yellow. Still there was no indication that there had been or was going to be any bloom.

I decided that they must be bluebead lily, so stressed out by the difficult environment that they didn't bloom, though it doesn't make sense to me that a fairly robust-looking plant doesn't bloom. Also, a difficult environment usually stimulates bloom. A mystery I will carry over to next Spring.

The cranberries were almost ready. Huckleberry had finished blooming and bore big, shiny, sweet black berries. This is the first time I sampled huckleberries, assuming they were only a poor cousin of the blueberry and cranberry, and the namesake of Huckleberry Finn. To my surprise, they were delicious by the handful, warm from the sun. If we had a bucket and more time, we should have been picking them.

We wandered into the bog farther than we had ever been before. Towards the rear were more woody shrubs and trees. The ground was becoming actual soil. This is what happens to a bog over many lifetimes. It gradually fills in and dries up and becomes forest.

Figure 64: Round-leafed sundew

We saw one very attractive shrub called false holly. It was six or seven feet high, with clusters of berries of a brilliant pink with a purplish bloom. I have seen this shrub in other places and learned that the berries are gone very soon. The birds eat them.

It was easier going in the bog this day than it had been earlier in the summer, and we could wander more freely. I explored the very edge where bog joined the trees, having read in the field guide that this was the environment preferred by some native orchids, but I found nothing.

Figure 65: Pitcher plant

I did find, though, down in a particularly wet spot, a patch of reddish looking moss, which, on closer inspection, turned out to be the sundew—a tiny, sticky-leaved insectivorous plant.

Then I found one pitcher plant.

There are two insect-eating plants of northern bogs. The sundew digests insects caught on the sticky hairs of its tiny round leaves. The pitcher plant, much larger and very unlike the sundew, also traps and digests insects, but uses a different strategy. When you find a pitcher plant, it is usually because of the tall, conspicuous red flower rising high above the moss. Look down, though, and you will find its tubular hollow leaves, filled with liquid. The watery liquid contains enzymes to digest insects that, guided by down-pointing hairs, take a one way trip into the drink. By digesting insects, both the sundew and the pitcher plant obtain the protein and nutrients they need to grow in the sterile sphagnum.

A curious fact is that mosquitoes are able to lay eggs and breed in the digestive liquid filling the pitcher plant "pitchers". This would seem to be impossible until you realize that mosquito larvae are swimming creatures that rise to the surface for air, and are not drowned in the liquid. Probably only dead, drowned insects can be digested.

Late in the Fall we returned to pick cranberries. Someone had beat us to them! I guess it could have been other pickers. The bog is right beside the trail and it would be surprising if there weren't others that had their eye on them. There had been plenty of berries though, and they were all gone. There was no evidence of trampling or bits of garbage lying around that would mean people.

We preferred to think that this was still our secret spot and that it had been animals or birds that had taken them. Mystery number two to investigate next year.

This bog I have been describing, near our cabin, is, of course, convenient and we can check on it frequently. It has a little of everything but, as bogs go, is rather small. Just a one pitcher plant bog.

The real powerful bogs are in Guysborough county, but we don't get there so often. Port Bickerton, however, is a must see.

To get to Port Bickerton from here, it is necessary to drive to Country Harbour, which is a very scenic drive past woods and lakes, and alongside the river. Then you catch the Country Harbour ferry, which is an exciting 15 minute trip across the harbour. A short drive past truly impressive bog land, and you are in Port Bickerton.

The destination is the lighthouse museum, which is reached by a roadway smack on the shore. At the museum you buy your tickets and get a fascinating history lesson and tour of the lighthouse from the nice lady inside. After that, eat your lunch at a picnic table or take your choice of several hiking trails departing from the museum.

We know where we are going. It is the trail the surfers take with their boards to get to Port Bickerton's highly-rated surfing beach. It is a long, rather narrow trail, though, and we have never met a surfer.

All along is handsome low, ericaceous vegetation with Labrador

Figure 66: Bakeapple

tea, blueberries, and sheep laurel in bloom, complemented by bayberry, low spruces and juniper. A few years ago when we were here, there were swarms of monarch butterflies. What they were doing in Port Bickerton I wouldn't know. There was no milkweed anywhere.

We aren't really going for the beach. Just before it is a long boardwalk through a really classic bog. Today, July 10, we've hit it right on. From the boardwalk, we see cotton grass as well as hundreds of pitcher plants and probably thousands of grass pink orchids.

Down in the moss is something else you don't see on our side of Nova Scotia—the bakeapple. That is what it is called in Newfoundland, anyway, where it is almost the national fruit. Here it is also called cloudberry. It is a small, very low plant with a nice white flower that ripens into a single fruit like a large yellow raspberry—to which it is related. It doesn't look at all like an apple but common names of plants don't have to make sense. Bakeapple is avidly gathered in Newfoundland, where it is widespread, and made into jams and sauces.

There are plenty of plants with ripening berries here but, in view of the protected nature of the bog, and the shame we would face if we were caught stepping off the boardwalk, we won't be back to pick. We are happy just to observe.

Figure 67: Dragon's-mouth orchid

Here I will mention another excellent boardwalk through a bog. This is one far away from Guysborough County, in the Cape Breton Highlands National Park. Mary and I visited it when we were in the park on June 26. It is off the highway passing through the park from Cheticamp, on the left just before Benjie's Lake.

As is everything else in the park, the boardwalk is beautifully built. Here we saw the cotton grass again, and there were dark pools of water where grew the curious buckbean and bladderworts. Neither were in bloom but everywhere else the colourful little *Arethusa*, or dragon's-mouth orchid, was blooming. It is only a few inches high and you must look very carefully to imagine a dragon's mouth. It is one of three species of showy pink or reddish early bog orchids found in Nova Scotia. The other two are the grass pink, which we had already seen, and the rose pogonia, which we were hoping to see.

We were excited to see the dragon's-mouth, although observing captive orchids is a tad less satisfying than finding wild ones. It is a beautiful boardwalk, and with its informative signs it makes an enjoyable experience for thousands of visitors every year.

We passed through Guysborough County once more, a week after Port Bickerton, on July 17. This was a driving trip with our daughters, Catherine and Margaret, through Country Harbour again, then following the 316 Highway to the end of the line, where it branches off to Canso in one direction and Guysborough in the other.

This highway takes you through Goldboro and then all along the Atlantic shore past Drum Head, Seal Harbour and Coddles Harbour, then to New Harbour, Tor Bay, and Larry's River. Across the water at Larry's River it keeps on to Charlos Cove, Cole Harbour, Port Felix and Whitehead, then to the Canso/Guysborough intersection.

A long drive but spectacular scenery all the way: the honest little French fishing villages and the Atlantic coastal shoals, islands and crashing surf. Hardly a store or gas station all the way.

Nameless bogs on both sides of the road tempted us to wade in and look around, something for which we unfortunately had neither the time nor the right clothing. If we had been able, though, I don't think we would have been disappointed. In most of them we could see pitcher plants with the naked eye, and when we did stop to use the binoculars, thousands of grass pink orchids. We'll go back again next summer, dressed to do some looking.

We'll be back before that if the weather gets hot in Antigonish. When the temperature is climbing to 30 where we live, it is less than an hour's drive to New Harbour or Tor Bay. You can feel the temperature drop by the time you are halfway there, and at the shore, you might need a sweater. We take a picnic and do some exploring, then return home in the evening fresh and rested.

The Fen

Figure 68: The fen

Classification of wetlands in Nova Scotia is based upon fertility and biodiversity, with swamps and tidal marshes at the top and sphagnum bogs at the bottom. The fen is just above the bog.

Wetland fertility is mostly a function of how much water moves through the system. Moving water carries nutrients, which boost fertility and productivity incrementally.

There is no moving water in the bog, hence no influx of nutrients. The fen is comparable but with the important difference that water in the fen has an outlet and slowly flows. Relative to the bog, this means more nutrients passing through and results in an entirely different type of vegetation.

The characteristic and defining plant of the fen is the sedge, which to most of us looks like grass. Sedges have their own particular beauty—some big ones with the allure of Egyptian papyrus. Sedges have triangular stems. Other than that, it takes an expert to tell them apart.

In the landscape, fens behave much like sphagnum bogs. Under wet and cool conditions, decomposition is slow and favours the formation of peat. For all intents and purposes, sedge peat is equivalent to moss peat and can accumulate in very deep deposits over time.

A fen often borders or encloses a permanent body of water, like a pond or a sluggish stream. This is the water feeding the fen with nutrients from the outside, making it a fen instead of a bog. Even so, decomposition is slow and fertility low.

Some insects such as the damsel fly and certain butterflies live in fens, as do salamanders, frogs and the endangered Blanding's turtle.

Botanically, learn to appreciate the sedges and grasses, which ripple in the wind and turn red and golden in the Fall.

The Marsh

A marsh in which the depth of water in summer is over 15 cm. is a deep marsh. A deep marsh may border or surround a lake or stream. Vegetation in a deep marsh includes emergents, also floating plants such as water lilies or duck weed, and submergents, which are plants that live completely underwater.

In the water and fertile mud of the freshwater marsh, plant growth is prolific and nourishes a vast and varied population of birds and animals. A Nova Scotia bulletin on wetlands puts it this way:

Marshes are one of the most biologically-productive types of wetland. Seasonal flooding continually adds nutrient-rich water and sediments to marshes.

Figure 69: The marsh

These nutrients nourish plants which, in turn, attract other wildlife. It is estimated that in terms of plant production, marshes are three times as productive as agricultural land and four times as productive as lakes and streams.

Figure 70: Swamp milkweed

It would seem that the marsh would also be at the top of the heap when it comes to capturing carbon dioxide, but it appears that carbon sequestration is almost the inverse of productivity. The sphagnum bog wins in sequestration because the carbon it captures winds up in deposits of peat and never gets away.

In the warm, wet and aerated conditions of the marsh, rapid decomposition keeps pace with growth. There is no appreciable build up of undecomposed plant material. For every molecule of carbon dioxide taken up in photosynthesis, one is given off in decomposition. I suppose that is what is meant by "carbon neutral".

The biological productivity of the marsh is seen in the abundant bird life—from sparrows and blackbirds to ducks and herons—many of which nest and raise their young there. Healthy populations of frogs and other amphibians, turtles, and fish attract the raccoon, the otter and the mink, their nocturnal wanderings written in tracks in the mud. Beaver and muskrat find a home in the marsh. In fact, many marshes are abandoned beaver ponds that have filled in over time.

The marsh is often bordered by shrubs, the red osier dogwood, the elderberry, the high-bush cranberry, and the wild rose among

Figure 71: Joe Pye weed

them. These are attractive flowering and berry-bearing bushes that provide food and nest sites for birds. Pink flowered swamp milkweed finds a niche here and attracts the monarch butterfly.

Typical of the marsh, too, are patches of the tall red-pink bloom-ing Joe Pye weed, the white turtlehead, and the blue-flag iris. (I read that Joe Pye was a native healer who used the plant to cure typhus fever.) You might see the tall purple fringed orchid. Grasses, rushes and sedges are a given. And it wouldn't be a marsh without cattails.

Figure 72: Purple loosestrife and cattails

You might see the attractive spires of purple loosestrife, an intro-duced Jezebel that is taking over wetlands and crowing out native that is taking over wetlands and crowding out native species across North America. Loosestrife has caused a lot of consterna-tion farther south and west of Nova Scotia, where it clogs up wet-lands but offers nothing useful to wildlife. Here, we mostly see them in roadside ditches, where they are pretty and probably harmless. The Nova Scotia Invasive Species Council is, however, keeping a close eye on them.

Aquatic plants form large mats at the edges of ponds and slow-moving water within the marsh, with roots in the fertile mud, leaves and flowers on top. The flowers of some are quite spectacu-lar—I would say exo-tic. They resemble tropical houseplants. They are so secure in the marsh, rooted in the mud and protected by the water, that they bloom abundantly, not having to get it all over with in a hurry.

The arrowhead, as its name implies, features large, arrowhead-shaped leaves that stand up a foot or so above the water. Flowering stalks support clusters of large, white- and yellow-centred, three-petalled flowers. Tubers of the arrowhead have been an important wild food source for indigenous people and are said to be good roasted.

The pickerel weed has large, oblong, pointed leaves that likewise stand up above the water. A showy spike of purple flowers arises from the base of the leaf. We have seen pickerel weed in bloom from the path bordering the duck sanctuary marsh in Brookfield, Colchester County.

Figure 73: Aquatic arrowhead

Figure 74: Pickerel weed and lilypads

The Flora tells us that it occurs in St. Andrews. That's where we live. This is embarrassing. We probably drive past it all the time, heading out to look for wildflowers farther away.

One aquatic plant we do know about here in St. Andrews is the wild calla, which grows at the edge of the pond beside the curling rink. This one really looks like a houseplant, with glossy, heart-shaped leaves rising half a foot above the water and its white hooded flowers, or spathes, just like the cultivated callas.

These three species with leaves standing up from the water are emergents. The water lilies are floaters. Like the emergents, they are rooted in underwater mud. Their leaves rise on elongated stems or

Figure 75: Wild calla

petioles from the bottom and terminate in the familiar pads that float and bob at the surface. Water lilies can grow in deeper water

Figure 76: Fragrant water lily

than emergents. Leaf petioles can be as long as seven or eight feet.

The floating leaves of the yellow pond lily, or cow lily, are large and oval. Flattened flower buds rise above the water and open to bright yellow buttercup-like flowers up to two and one-half inches across. The species is closely related to the lotus of ancient times, but falls short in the fragrance department. The flowers are said to have a "sickly" smell.

Our other waterlily smells wonderful. For this reason, it is known as the fragrant water lily. This is the classic water lily with its floating "pads" for frogs to sit on. The lovely white flowers scent the air on sunny days, and resemble peonies floating on the water.

The Swamp

Figure 77: The swamp in summer

The swamp is much like the marsh but drier, with bushes and trees growing in it. As in the marsh, water is moving and transporting nutrients; but flow is seasonal and shallow, allowing woody plants to grow in the higher spots. Like the marsh, the swamp is biologically productive, supporting significant plant, bird and animal life.

Mosses and ferns are plentiful. There are usually nice patches of bunchberry and creepers like veronica, partridge berry and twin-flower. There may be pink lady's slippers and other orchids. There will be wild lily of the valley and bluebead lily throughout.

One notable discovery I made in a swamp was a large witch hazel bush in full bloom in November. This is the witch hazel's moment, and is the last bloom of the season. The spidery yellow flowers open at the same time as the leaves are turning yellow and dropping off.

Figure 78: Clintonia-bluebead lily

Entering a swamp is entering a very wild place. The untended swamp (and I doubt if there are tended ones) is not very easy to explore. By definition it is wet. It is lumpy. and may be a morass of fallen trees with roots sticking high into the air. Exploration is arduous and slow and not often tempting, unless you are hunting or trapping or hoping to find a rare orchid.

As do all wetlands, swamps play an important role in the regulation of water flow. Large areas of swamp are able to disperse and tame dangerous flooding while, on the other hand, releasing water during dry periods. It is unfortunate for swamps that they are drier than marshes and often the chosen target of draining and filling for development.

That is what people do. Beavers develop swamps in their own way, and seem to think that they were here first. If the water making its way through the swamp has an outlet that can be dammed, that's what they do. Water backs up and floods the swamp, regardless of the plans people have for it. Then you have your classic beaver pond, with lodge and dam.

Beavers gnaw and fell trees outside the pond, cut them up and drag them in for food (the bark and the soft layer of wood just under it) and for building supplies. Poplar is their favourite.

Soon trees in the flooded area drown, die, and stand like scarecrows until they crumble and fall. Woodpeckers, owls and wood ducks, flying squirrels and other birds and animals inhabit them while they stand. Sun floods in where all was shady before.

Sooner or later, the beavers run out of trees to drop, and move on. The dam falls into ruin, and the pond disappears. With trees gone and abundant sunlight, grasses and other vegetation quickly take over. A swamp has become a marsh, and life goes on.

These days, many people are working in the woods, housing developments proliferate, and a beaver has to be careful where he puts his pond. Beavers' plans and people's plans come more frequently into conflict, with humans inevitably the winner—at least in the short term.

At one time we owned a farm that included a large area of swampy forest. One morning we woke up to notice that we seemed to have a lake. We rubbed our eyes and it was still there. Sure enough, beavers had dammed up water flowing out of the swamp.

This was quite exciting. We had often talked about digging a small pond, but this was a big pond, and we got it for nothing.

The beavers had made one mistake, though, a big one. The water flowed out of the swamp through a culvert under a woods road. The beavers had blocked up the culvert and built their dam half on the road.

This didn't sit well with certain people. Trucks couldn't get by. In came the excavator and tore up the dam. The national animal moved on to try somewhere else.

When the dam was torn up, though, there was a worrisome discovery. The beavers had found a sheet of plywood somewhere and worked it into the dam. Dam building technology had taken a quantum leap. If beavers get their hands on more plywood, they may build dams we can't break.

We were sad to lose our lake. Unfortunately it had been there long enough to drown and kill the trees. We lost our lake and our forest. The poor dead trees, roots losing their grip in saturated soil, blew down in a tangle in a big wind storm.

The area is still wet, and now that the trees are gone, flooded with sunlight. The swamp will go to marsh.

Lakeshores

Figure 79: The lakeshore

The freshwater lakeshore is that ribbon around the lake where water meets land, but is neither one nor the other. Depending upon the slope of the land, the nature of the lake bottom and the shore, and many other factors, the lakeshore environment can take many forms.

No two lakes are alike. Some are man-made, and some are natural. Some are deep and some are shallow. Some are big and some are small. Some are surrounded by towering trees and others by bogs. Some have an inlet and an outlet, some don't. The character of a lake is determined by all these factors, and more, as is the vegetation inhabiting their shores.

Shallow lakeshores tend to be boggy or marshlike. Frequently, they are indeed an extension of a marsh bordering the lake. The vegetation is much like that of the marsh—sedges, grasses, rushes, and ferns, with water lilies and other floating plants, and underwater submergents where the water becomes deeper. Of the thousands of lakes in Nova Scotia, a few, at least, have gone down in botanical history.

A hundred years ago, the wild and sometimes unnamed lakes along the Tusket River and its tributaries interested Dr. Fernald and his team. Here, in Southwest Nova Scotia, winters were mild and foggy and those species that had come over the prehistoric land bridge from the south could survive. Here, again, were the curly-grass fern, the inkberry holly, the pink coreopsis and the most elu-

Figure 80: Plymouth Gentian

sive and beautiful of them all, the Plymouth Gentian. Dr. Fernald:

> Our route lay up the Tusket valley and, after a few stops, we succeeded in getting above Tusket Falls, when some one thought he saw an interesting plant on a wooded slope

above Tusket (or Vaughan) Lake. The shore of the lake was obviously of no interest, being bushed close down to the water and with absolutely no beach exposed, but, tiring of waiting for the others to return, I pushed idly through the bushes to the water's edge and there, with flowers fully expanded under several inches of water, was the beautiful Plymouth Gentian.

The range of the Plymouth Gentian and other lakeshore rarities was restricted to only a handful of lakes in Fernald's time. Their survival is even more precarious now.

A survey of endangered lakeshore species in the Tusket River watershed was conducted by Paul A. Keddy in 1985. He concludes:

Older records (Fernald, 1921, 1922; Roland and Smith, 1969) indicate that these species were always rare in Nova Scotia, but habitat loss from human activity is now a growing threat. Raynards, Vaughan, Gavels and Kings Lakes are now hydroelectric reservoirs, and species such as *Sabatia kennedyana* (Plymouth Gentian) and *Coreopsis rosea* have apparently disappeared from them. Bennetts, Wilsons, Kegeshook and Pearl Lakes are all being developed for cottages. Cottages not only lead to trampling, but vegetation is sometimes deliberately eliminated to provide swimming areas. All-terrain vehicles are causing increasing damage, and several *S. kennedyana* populations seen in 1982 were heavily damaged.

This is a sad but not uncommon story and shows the plight of native vegetation and wildflowers wherever people and flowers collide.

Another rare lakeshore plant, the water pennywort, rumoured to exist in Nova Scotia, was eventually discovered in 1921 at Wilson's Lake by that same intrepid botanist, M.L. Fernald. The water pennywort was not seen again for fifty years, until Albert Roland found it at various places on Kejimkujik Lake. Since then, though rare, it has been found at other locations not far away.

When Mary and I camped at Kejimkujik a couple of years ago, there were signs at every campsite warning campers and boaters to watch out for and not damage the pennywort. In between hiking and canoeing, we looked for it but didn't find any. In truth, it is very rare—found only in a few hidden bays of Kejimkujik and a couple of hard-to-get-to lakes nearby. Most visitors to the park, like us, probably never see it.

Figure 81: Water pennywort

In October one year, we were canoeing on small lakes in the To-beatic Wilderness, not far from Kejimkujik. These lakes were nicely forested down to the water's edge, and at the edge of each was a fringe of royal fern in bright yellow and orange fall colour. I had been to the same lake in summer and hadn't even noticed these ferns.

They were doubly noticeable this trip because the lakes were mirror-smooth and the brightly-coloured ferns were reflected perfectly in the water, upside down. The forest growing to the water's edge was likewise reflected perfectly. It was possible to gaze deeply into the forest as we paddled along by simply looking into the water.

Water levels were low at the time, and there were rounded granite boulders jutting above the surface. These, too, were perfectly reflected and gave the bizarre impression of granite balls floating in the lake. I took a picture of our friends in another canoe, and the water was so still, and the reflection so perfect, that when I had the film developed and printed, it was impossible to tell which way was up. We finally figured it out only because there was a tiny piece of floating wood that was in the sky if we turned it wrong side up.

Besides the unbelievable clarity and stillness of the water, the things I remember most about that trip were the handsome expanses of bearberry; the discovery of cat brier, which is viciously thorny and runs horizontally through the forest like green barbed wire; and the bear hunters in tree stands baiting the bears with day-old doughnuts from Tim Horton's.

In Summer, Mary and I go to lakes more for recreation than for botany, but we always check to see what grows. Last July (2022), we rented a cottage on a lake in Maine, where we met up with our children. Our time was spent canoeing and swimming and fishing off the dock when we weren't sunk into the lawn chairs.

We did notice the lakeshore was shallow and reedy. In deeper water was a floating plant we had never seen before. It had round leaves with a stalk that attached on the bottom right in the centre. It looked like maybe a small water lily, but we eventually discovered small insignificant reddish flowers that stood up on stalks above the water.

When we canoed, we paddled through it. When we swam, we swam through it, pulling stems off our arms. When we fished, pieces of it got caught on our hooks and we pulled it in.

When we lifted it out of the water, it was a bit of a shock. The long stems and leaf petioles were coated with thick, clear, slippery goo. At first, I thought it was a disease of some sort, but every bit was coated.

A floating plant covered in slime wasn't too hard to find in the wildflower guide and we identified it as watershield. It occurs commonly in Nova Scotia as well, according to the Flora, but we have never seen it. We learned that is indeed in the same family as the elegant water lily but on the losing end of the genetic crap-shoot and destined to be plain and slimy.

Another lake we visited last summer was Benjie's Lake in the

Figure 82: Bruce and Mary at Benjie's Lake

Cape Breton highlands. This is a smallish, picturesque lake sur-rounded by mossy spruce woods, and much different than the other lakes we visited.

Benjie's is reached by an easy 1.5 km. trail off the Cabot Trail at the top of French Mountain, and this time we were looking at plants. The park map described it as boreal forest, which at that elevation means spruce and fir, though we hoped that we might get glimpses of barrens as well.

In July, at the time of year we were there, the Cape Breton Highlands park is rather crowded with people, and a lot of them were heading to Benjie's Lake. The first part of the trail is straight, and from the car it looked like people lined up to get into a movie. We took our place in line and before long hikers were spread out and it didn't seem crowded anymore.

The first wildflower we saw of interest was the Canadian burnet, with its curious bottle-brush-like spikes of white flowers. There was lots of it just at the start of the trail. It is something that is mostly only found in northern Cape Breton.

We worked our way along towards Benjie's, checking the sides of the trail as we went, looking for anything interesting. By that I mean anything we weren't expecting to find or which we couldn't identify.

Everything along the trail was interesting. It was familiar stuff but, as always, arranged in a multitude of inspired combinations.

Figure 83: Canadian burnet

Mary and I worked most of our lives in landscape design and landscaping, and did our utmost to create beautiful plantings—often using the same plants we were seeing on the way to Benjie's. Somehow, our work inevitably fell short of that of the master.

We were up high, and off into the distance were hillsides of dark spruce and fir, interspersed with grassy meadows, of which I will have more to say. The trail rose a bit, and fell, and wound around, and we scarcely glimpsed other hikers.

As we poked along, or "sauntered", as Henry David Thoreau would put it, we were certainly the slowest hikers on the trail, and were passed by others, including runners with earbuds in their ears, which makes you wonder what part nature really plays in their lives.

Before too long we could see the shining waters of the lake through the trees. At the shore, the park people had built a nice viewing platform and a bench. With our binoculars we could see ducks out on the water.

The whole setting was very scenic, but it didn't look too easy to explore the shore, which was boggy in one direction and over-grown with bushes in the other. We decided to sit and think, and eat the lunch we had brought along.

After lunch we tried the shore, but were soon mired in mud and blocked by the outlet stream. To the left, though, was what seemed to be the start of another trail, with boards laid over the mud. We decided to give that one a chance.

Once we were across the boards, the trail was drier and quite easy to follow. By the sections of old boardwalk thrown off to the side, we concluded that this must be a piece of abandoned trail. It was fairly dry now but must get very soft in the wet months. Apparently this section of the trail had been relocated farther uphill.

An abandoned trail has a certain allure. We sometimes had to skirt around or climb over fallen trees. From the sides of the trail, bunchberry, mayflower, blue-bead lily, ferns and the like had crept in and almost covered the old path. We suspected that there were probably lady's slippers and maybe other orchids in season, but by this point they had died down and disappeared.

Human footprints were nonexistent, but in every muddy spot were the tracks of moose. The park trail had become a moose trail. We followed along until it did indeed come out on the main trail, then we hiked back to the car.

It was still early afternoon, and the trail to Benjie's Lake hadn't taken too much out of us, so we thought we might try the nearby, world-famous Skyline Trail. This trail had been written up in *Na-*

tional Geographic magazine for its famous look-offs above the crashing ocean far below, and for the moose.

Moose were so plentiful in this part of the highlands that they were a signature feature of the park. Years before, I had hiked this trail with a friend and we did see several moose. At one spot, a cow moose and her calf were parked a few feet off the edge of the trail. At this distance, she looked as big as a draft horse and capable of inflicting serious damage.

We had already hiked several kilometres of trail and were closer to the finish than we were to the start. The question was whether to turn around and go back the way we came, or to attempt to edge by the moose and keep going. It seemed very likely that she might attack us to protect her calf.

In the end we did very gingerly creep by the moose, who paid us absolutely no attention, and we finished the trail.

Mary and I had driven by the entrance to the Skyline a few times, but avoided it because the parking lot was as full of cars as the one at Walmart. Today, though, we resolved to do it anyway.

The crowd was unbelievable. People of every age and race. Many senior citizens of various degrees of physical fitness, some limping and walking with a cane. There were couples with small children on their backs or in a stroller, or just tagging along. There were state-of-the-art-equipped trekkers and eager first-timers with completely inappropriate clothing and footwear.

But it was a good day, and a party atmosphere. Everyone was smiling and happy. All were ready to hike that trail all the way.

The trail is well-built, but it was a long one, and steep in places. Much of it was boardwalk. Mary and I were running late so we hurried along.

Along the way were informative signs about botany and geology and moose. There were signs about how to behave if you encountered a moose. There were signs describing the damage the overpopulation of moose were doing to the forest.

Moose are devouring tree seedlings as fast as they come up. They also eat the lower branches of the grown trees as high up as they can reach—which appears to be at least eight feet. Because there is no forest regeneration, trees die out and what was once a forest becomes a meadow. Hence those attractive meadows we saw in the distance on the way to Benjie's Lake. We could see many more from the Skyline Trail.

There was a huge fenced enclosure beside the trail. A sign said

Figure 84: The view from the Skyline Trail

that the enclosure was to keep moose out so foresters could study how the forest regenerates without moose.

The trail we chose was 6.5 km. return. Long enough. At the end of the trail were steps leading down to one viewing platform after another.

The view from each platform was breathtaking. There were no trees here—mostly rock—and the cliffs fell away to the blue ocean far below. This was a true alpine environment.

There were signs describing the rare and endangered golden heather, which is found here and almost nowhere else in northern Nova Scotia. At the right season it has big yellow flowers, but it was not in bloom when we were there. We did finally locate it down in the mat of blueberry and bearberry, and saw more after we learned to recognize the foliage.

From the Cabot Trail far below, this enormous promontory of rock appears to be covered with grass, but it is actually those mats of blueberry, bearberry and golden heather, interspersed with vast sweeps of blue-green juniper.

By now, the day was getting on and we turned back. Numbers of the young families, the old people, the lame, the inappropriately dressed, that we had seen at the start had made it all the way and were on their way back, too—still smiling. In fact I didn't notice anyone who had turned around part way. And there were still more coming.

Maybe it would be different in bad weather, but I will remember the high spirits of the people we saw that day, and their joy in being outdoors and hiking. It makes us appreciate the vast benefit to humanity, of parks like this.

We never saw a moose.

Wilderness at home

We have talked about exploring in bogs, which can be done with care. Swamps can be handled. Fens, marshes, and lakeshores are a different matter. Their very nature, bordering bodies of water of indeterminate depth, makes them treacherous to step into. One is almost certain to go in over the boot tops or worse. Disaster is too often hidden by innocent clumps of vegetation looking solid.

The best and almost the only way to proceed is by kayak or canoe. This can be an enjoyable experience in many ways. It is a smooth and silent way to approach all forms of wildlife, from muskrats to moose, bitterns, herons, loons, and ducks of all kinds with their young. Shores can be examined from the inside, avoiding rose bush scratches, twigs in the eye, and tumbles into murky water. It is just about the only way to go, unless you are lucky enough to have wetlands just off your back deck like Mary and I do.

Our house sits on a rise, with big maples and ash trees. In front is the road. In back the land drops sharply down across an old barbed-wire fence into what was once a soggy cow pasture.

A small brook runs along our property line and under the road down to the South River beyond. The brook is fed by overflow from our neighbour's pond, spreading out through cattails and tall swamp grasses before flowing away. In the Spring, it is joined in wet weather by runoff along the fence line from an alder swamp above.

Beyond the alder swamp, the land rises steeply again past hay fields and patches of woods into the distance. All of this we can watch from our deck on the sunny back side of the house.

So, you see, we have a little marsh, a little swamp, and what passes for a fen when water is flowing along the fence line—a veritable nature park just beyond our deck chairs.

There is not a lot of unusual plant life to be seen. The cattails are, of course, to be expected in the marsh and are attractive when they are green and growing and when they mature with their "tails". The grasses along the fence and between the marsh and the alders grow thick and tall and wave in the wind. Here and there are struggling elderberry, swamp rose, and steeplebush spirea. Some summers I see a solitary spike of purple fringed orchid.

Making up for the uniformity of plant life, though, is the panorama of creatures that crawl, swim, or fly.

Bird life is the most animated by far, starting with the red-winged blackbird trilling from the cattails. All manner of sparrows, warblers and finches feed and nest in the alders. In the sky, the little blackbirds chase away the grackles, the grackles chase away the crows, and all three chase the hawks and eagles. Ducks some-times splash down in small patches of open water. We have seen snipe, woodcock, and ruffed grouse. It is a very easy way to watch birds, seated in chairs with binoculars.

Other wildlife we have seen include a snapping turtle working his way up the brook towards the pond, squabbling muskrats, skunk, red fox and deer in the fields up above. A couple of sum-mers ago our neighbours told us that a bear had run through the swamp, but we weren't home to see it.

In May, the symphony of frogs and peepers makes it difficult to sleep; and on damp, warm nights in June, the marsh twinkles like the stars in the sky with thousands of blinking, spiralling fireflies. A magical place.

On the other side of the house, with summer coming on, there will be magic, too. Along the roadside, in the ditches and run-out hay fields, comes a parade of familiar summer flowers, the illegal immigrants.

6: Roadside flowers

Figure 85: Dandelions in the pasture

It begins with the coltsfoot, then the dandelions, followed by but-tercups, daisies, purple vetch, hawkweed, chickweed, red clover, lupin and yarrow for a start. Say the word wildflower, and these are probably the flowers that come to mind.

Later on, we can add chicory, evening primrose, Queen Anne's lace, black-eyed Susan and fireweed. The familiar flowers we see from the car.

It may surprise you that none of them are native to Nova Scotia. All of them introduced.

But let's look at it this way: every one is native somewhere. Most are said to have come from Eurasia, which can be anywhere in temperate Europe or Asia. It could be red clover that entered illegally in a bag of oats, or daisies smuggled across the border by an immigrant woman to remind her of home. It could be the creeping buttercup coming off the boat in a sack of muddy potatoes, or vetch or trefoil imported by big shots for cattle fodder. Dandelion fluff could have blown across the border from anywhere.

None of these species were unhappy at home or had any desire to be noxious weeds. Most came in good faith with the pioneers, and I think by now we can give them Canadian citizenship. In gratitude, they get together in June, July and August and put on a colour pageant to gladden our hearts.

Early in May, when along the road the coltsfoot is finished and tufts of grass are greening up on the sunny side of the ditch, out come the dandelions. Farmers' fields, now that herbicides are frowned upon, become dandelion carpets of golden yellow like

Figure 86: Buttercups in the ditch

something out of *The Wizard of Oz*. Later, their leaves are made into hay with the grass, and cows and horses eat them.

When dandelions wane and go to seed, act two begins, directed by the creator, the master landscaper. In the ditch, the grasses are getting high, and in some, cattails are growing.

As Spring heads into Summer, and we are not paying attention, masses of shining buttercups appear out of nowhere. This is the creeping buttercup. There is a tall buttercup, too, and a few of these go in for height. Clumps of daisy bloom in strategic places. Red clover adds globes of red to the mix, and purple vetch is directed to climb the grasses. With a final flourish, chickweed is scattered throughout to tie everything together with its tiny white flowers like baby's breath.

The Master never does the same planting twice. Every roadside ditch is another work of art, and warrants slowing down for a better look. The beauty and originality of each composition is such that one hardly notices the beer cans and chip bags.

The edge of the road, on the one side of the ditch, and the farmers' fields on the other side, are not forgotten.

The edge of the road is a tough, dry environment, a far cry from the muddy ditch. By Summer, roadside grasses are stunted and dried and scarcely alive. Clumps of rabbit-foot and white clover appear. These clovers are members of the large legume family, all of which have the ability to transport or "fix" nitrogen from the air directly into the soil by means of symbiotic bacterial nodules on their roots. This bit of nitrogen, so essential to plants, enables clumps of daisies and patches of hawkweeds, the orange and the yellow King-devil, to join the clovers and thin grass on the sun-baked shoulder of the road.

Hawkweeds, with their vividly-coloured, dandelion-like flowers, are the toughest weeds going and will grow just about anywhere. The yellow one must have done something naughty to earn the name King-devil.

Both the yellow and the orange form colonies, and often occur together. The Master puts them in drier spots where the buttercup won't grow, and across the ditch in old pastures and hayfields, which at this time of year need some colour.

Old farmland ready to become wildflower meadows was formerly more common than it is now. These are the worked-out and abandoned fields along the road, extending from sagging barbed wire fences to broken down barns and empty farmhouses. These forgotten pastures and hayfields slowly fill in, with young white spruce and alders poking up here and there.

Figure 87: Yarrow, purple vetch, and yellow bird's foot trefoil

The fields will become forest again one day, as have most of their kind; but until then, they are ready for beautification. Again, daisies, red clover, purple vetch and hawkweeds are the big players, with some pink alsike clover, yarrow, tall buttercup, toadflax, chickweed and eyebright thrown in. The tenacious timothy and fescue grasses play their part by supporting the flowers, while legumes, the vetch and the clovers, play theirs by "fixing" nitrogen to feed the cast.

The finished composition is inevitably tasteful, dynamic, and ever-changing. Its beauty puts to shame the phony picture on the box of wildflower seed that never grows.

It will be back next year for free.

Mary says the most spectacular of these old-field meadows that she has seen were on the Magdalen Islands, where they enjoy the full sun and sea air.

Figure 88: Daisy and yellow and orange hawkweed in the hayfield

Incidentally, these worn-out pastures and hayfields are a refuge for wild species, among them the threatened songbird, the bobolink,

Figure 89: Black-eyed Susan

which nests in the grass. The highly fertilized and tended fields on today's modern farms, while admirable, are mowed too often to al-

133

low the bobolink to raise its young. Other than the spring show of dandelions, they are not very welcoming to wildflowers, either.

As summer moves on, the roadside flower show continues to evolve. By now the lupines have run their course on the slopes, and are followed by black-eyed Susan and Queen Anne's lace. All along the road, tufts of brilliant yellow bird's foot trefoil—another legume—compensate for faded buttercups, and patches of soft fox-tail grass wave in the breeze.

On banks, stalks of mahogany-coloured dock, soft yellow spires of evening primrose, mullein with its velvety leaves, and sprays of sky-blue chicory overtop the clover and mauve flowered knapweed —a curious plant like a thornless thistle. Bright pink fireweed brightens old clearcuts, and pearly everlasting is ready to cut.

The roadside slides colourfully into autumn in the care of a couple of Canadians, the goldenrod and aster.

The come-from-aways have done their show, and given us more than our money's worth. We won't come down too hard on them if they stray into our flowerbeds and gardens from time to time.

Figure 90: Knapweed

7: Summer in the forest

In the forest, now that it is good and warm, trees are leafed out, ephemerals are finished, and there is a brand new floral show underway in the green filtered light. By mid-July there is no better place to be than under a tree. Summer plants of the forest are adapted to and happy in the shade.

Before I go any farther, I must pay tribute to a handful of hard-working native plants, which are so common and widespread that we sometimes don't see them, but which make our forests and wild

Figure 91: Summer in the forest

places what they are. To set the stage, let us first thank the mosses and ferns.

No matter where you wander in the wild outdoors, you will find mosses and ferns. Mosses grow in nearly any environment. They have the amazing ability to dry up to a crisp in dry weather, then soften up and come alive again when it rains. We take them for granted and only a bryologist can tell them apart, but they add im-measurably to the landscape. In the cool weather of Spring and Fall they are particularly luxuriant, glowing with different shades of orange, green and even red.

Figure 92: Mosses and ferns

The close companion of the moss is the fern, also often taken for granted. You will find clumps of wood ferns and patches of New

York fern where land is somewhat damp and shady, hay-scented fern and bracken where it is drier. Many other ferns join in, depending upon the levels of shade and moisture.

Figure 93: Wild lily-of-the-valley

In with the moss and ferns, to complete the composition, are the flowering plants. Four of these, the wild lily-of-the-valley, the blue-bead lily, the bunchberry, and the wild sarsaparilla, do most of the work. They may be found in any environment, from forest, to gypsum cliff, to bog and, more often than not, all grow together. You are sure to see them wherever you wander in the wild.

The wild lily-of-the-valley looks superficially like a miniature version of the tough, imported European lily-of-the-valley. It is a very pretty plant in its own right and needn't be compared to the import. It has another name, Canada mayflower, which is good, unless it is confused with the actual mayflower. I don't think the plant is sensitive about its name anyway, so I will apologize and continue to use the name wild lily-of-the-valley.

We don't need the tame lily-of-the-valley, because the wild one, or Canada mayflower if you prefer, grows prolifically, with starry white flowers in spring and red berries in fall. Our wild places wouldn't be the same without it.

Where wild lily-of-the-valley grows, you will probably find the bluebead lily as well. Bluebead lily, or *Clintonia*, is widespread and undemanding. All it asks for is a bit of unspoiled ground, be it bog

Figure 94: Clintonia, or bluebead lily

or swamp or forest. It is omnipresent in the wild and easy to take for granted.

In early June it blooms with understated yellow bells. By end of summer, the surprisingly-beautiful blue berries, the beads, are ripe. They are big, shiny, a lustrous sky blue colour, and said to be toxic, which adds to their mystique.

The plant tends to grow in colonies, among wild-lily-of-the-valley and bunchberry. Its straplike leaves are a light glossy green.

The bunchberry is the third of the big four. Quite different in appearance from the first two, it is just as widespread, and the three are often found together. If the bunchberry was rare, it would be a celebrated wildflower. It is a herbaceous member of the dogwood family, which is mostly trees and shrubs, and it grows in patches over the ground, sometimes very extensive ones. In common with the woody dogwoods, it has shiny strong-veined leaves that turn to a rich red in fall, and a rather large, snow-white flower in spring, which isn't a flower at all.

Figure 95: Bunchberry

What appear to be the four white petals of the flower are something called bracts. Because these bracts are not flower petals, they last a long time and don't fall off. The actual flowers are a yellowish cluster forming what we would call the centre of the daisy, if it was a daisy. By fall, the bracts have turned brown, and taking their place is a cluster of vivid, red berries, the "bunch".

Again, the bunchberry grows just about everywhere, though not so much in dry ground, and you will surely see it often.

The wild sarsaparilla is the fourth plant you are sure to see. It is taller than the other three and comes up strongly first thing in the spring, blooming shortly after with a ball of greenish flowers. When it is first bursting through the ground it looks like it could be something exotic and gets us excited. As soon as we see the flowers, though, we know what it is and let out our breath. It is in the same family as, and could be mistaken for ginseng, though that sort of ginseng doesn't occur in Nova Scotia.

By midsummer in the forest, it can almost be the dominant

plant, growing about a foot and a half high. In fall, it has a round cluster of black berries.

These four herbaceous species, plus the mosses and ferns, are the backbone of many wild plant communities. Their appearance is variable, depending upon where they are growing. In rich ground, they grow their best.

Figure 96: Wild sarsaparilla. with bog laurel in bloom

The little, wild lily-of-the-valley can be found with leaves nearly as big as the tame one. Where it pokes through the moss in a bog, it may be only an inch or two high.

The other species are likewise more luxuriant in better soils, but all will cope with any piece of wild land. In many places, these four species alone, along with the mosses and ferns, essentially comprise the herbaceous landscape.

We plant enthusiasts search for botanical rarities, but thanks to these four and the ferns and mosses, most who venture into the outdoors wouldn't notice anything missing.

The combination of mosses, ferns, and flowers in the forest varies with tree species, soil and shade There are different sorts of forest in Nova Scotia, depending on many factors, and different sorts of shade. Each type of forest has its characteristic understory companions, there is much overlap, and one grades into another. To complicate matters further, our forests have been under siege for almost three centuries.

As things stand today, less than 1% of our forest is over a hundred years old. Over one hundred years is considered old growth forest, but does not imply virgin forest that has never been touched. If there is virgin forest left in Nova Scotia anywhere, it is probably skinny bog black spruce with zero commercial potential. It is also possible that there is forgotten forest in ravines or on hillsides so steep that they were never logged.

Where there is money to be made, though, loggers figure out how to get the wood out. Just look at British Columbia. Trees the size of railroad cars felled with the axe and slithered down the mountain using primitive steam engines, chains and winches. Today they use helicopters.

Despite the hype of "Nova Scotia Needs Forestry", the forest industry in this province today is less than the shadow of a shadow of what it was. As early as the mid-1700s, the great powers (England and France), observed with glee the virgin forest and the trusting and poorly armed native population, and began to help themselves to the best. At first it was giant timbers squared with the axe and sent across the ocean, then masts for the navy. The English, who, in the heyday of sailing ships were building the world's most powerful navy, had run out of timber.

Access to the pines of New England ended with the American Revolution of 1776. Their sights turned to Nova Scotia. In 1774, in anticipation of trouble in New England, the King's government in London declared the entire island of Cape Breton to be His Majesty's Timber Reserve.

Not every tree was suitable for a mast. The huge wooden warships and freighters of the day demanded the best. A good mast had to be tall, straight, of sufficient diameter, and free of any rot or defect. Only the native white pine met these requirements. A mainmast could be 100 feet tall and three feet in diameter at the butt.

Suitable trees were marked and felled. Untold numbers were shipped out of Nova Scotia during the next hundred years. I wonder if there is a single mainmast pine still standing in Nova Scotia in 2023.

By the 1800s, the Maritime provinces were becoming rapidly settled, cities and towns were growing, and logging and sawmilling reached a fever pitch. We were a society that used wood for everything. There was a sawmill on every river. Wooden ships and boats were built in every harbour around the coast. Every item of everyday life was fashioned from wood from our forests. Nothing except tea came from China.

In his book *Great Forests and Mighty Men*, David Lee puts it in perspective:

> The nineteenth century was the great age of expansion in the Western world. It was the age of the Industrial Revolution and the migration of millions of Europeans to North America. The nineteenth century was also the age of wood, for it is clear that neither of these momentous historical developments would have been possible without this all-purpose building material. Strong, durable, flexible, versatile plentiful, and cheap, wood was employed to make a panoply of goods. It was used to provide people with small wares like furniture, hand tools, and farm ploughs as well as larger assets such as houses and barns. It was used to construct the wharves and ships that carried people to the New World and build the cities they founded. It was used for the telephone poles, railway ties and railcars that connected them. It was used to build the factories of the Industrial Revolution as well as some of the machinery within them. It was used to make the millions of boxes, crates, casks, barrels, kegs, and hogsheads needed every year to hold the products of this revolution. It was used to manufacture the wagons and wagon wheels that carried those products to market. Finally, in an age of growing literacy, it furnished the fibre to make paper for the burgeoning book and newspaper trade.

Nova Scotia was a busy place in the days before our fathers were born, and the forests were endless, or were believed to be. Thousands of men spent the winter months in camps away from their boats or farms, cutting and hauling trees for lumber companies. These companies housed and fed the men in rough log shanties.

Logs were taken out by teams of oxen or horses pulling huge loads over the snow and ice. Vast piles of logs were stacked beside lakes and rivers waiting for the ice to break up in Spring, when they were "driven" downstream to the mill.

Most lumber companies had a sawmill, selling and exporting lumber and making their owners rich. Some were so prosperous that they built towns deep in the woods for their workers and even railroads to bring out the logs.

The railroad locomotives were fuelled by burning wood or coal and started fires along the tracks. There were no fire-fighting crews or water bombers in those days. Fires burned until they went out on their own. They might burn huge tracts of timber, but no one cared. There was plenty more.

Since those early days, technology has drastically increased the efficiency of cutting down trees. First, the power saw and 4-wheel-drive forwarder, finally the giant hydraulic harvester, with which one operator can cut, limb, and haul out trees twenty-four hours a day.

Equipment got bigger, jobs got scarcer, trees got smaller and look at our forest industry today. Sawmills are mostly gone and trucks haul nothing but pulpwood and matchstick-sized poles for biomass.

To be truthful, once immigration to Nova Scotia and colonization really began in the 1700s, clearing the forests was both necessary and inevitable. Early settlers arriving in Nova Scotia, such as the destitute Scottish families disembarking in Pictou after a hellish eleven week journey on the rotting Dutch scow, the *Hector*, needed desperately to build shelters and get crops in the ground. The towering forest stood in the way.

In his book *Forests of Nova Scotia*, Ralph S Johnson describes their plight:

It is hard today to imagine the difficulties that greeted the white settlers in the eighteenth century. They arrived by ship at places that appeared forbidding and offered only danger, hardship and deprivation. Few were prepared for life in the new land. Besides the Loyalists and disbanded soldiers, some settlers came because of persecution and prejudice in their homeland, some as a result of fabulous promises by their governments, some in search of adventure.

Some brought building materials and furniture with them, but most arrived with only a few crude tools, cooking utensils, their clothing and food supplies to last only a short time. For the most part they depended on the forest to provide fur, clothing, logs, bark and moss to build shelters and wood for fuel. Collecting these materials and building shelters with the equipment at hand was gruelling work. Many had no horses or oxen and had to drag the logs, one at a time, by means of rope, with the front end of the log riding on a low, roughly built sledge.

Somehow these settlers hung on and pushed back the forest for farms and villages. With the flood of immigrants arriving in Nova Scotia in the 1800s came horses, oxen, and manpower. Soon every stream had a sawmill and the log drives began. Railroads were built to take out the logs, and schooners transported sawn lumber all over the world.

The best farmland in the province was quickly taken up by those who knew farmland, and the luxuriant hardwood forest cleared. Often the timber was more than could be used, and felled trees and stumps were simply burned.

Where the forest stood, we have traded it for beautiful cities and towns, picturesque pastures and prosperous farms.

It might seem that our forests are gone, but that is not completely true. Many farms include extensive woodlots, or treed banks along rivers. Around most cities and towns there are parks and pieces of land where the forest is reasonably intact. The national and provincial parks and land trusts are preserving large areas of forest. And finally, in a province as stony, boggy, steep and cut with canyons and ravines as this one, you can't get 'em all.

Maybe only one percent of our forest is old forest, but considering the acreage of forested land in the province, one percent is enough for plant hunting. We will have to be content.

Furthermore, the forest doesn't have to be "old" to be interesting. Clearcuts are out, but woods that have been selectively or only partially logged often have their understory intact. Forests no longer stretch from horizon to horizon, but sizable remnants are not hard to find, and offer good plant hunting.

In his *Flora of Nova Scotia*, Albert Roland describes three major forest regimes, the Boreal element, the Canadian element, and the Alleghenian element. Each is defined by the prevalence of certain tree species and associated flora.

The boreal forest is primarily evergreen spruces, pine, larch and fir, with some birch and poplar thrown in, and occupies the most

Figure 97: Boreal forest

northerly and extreme environments in the province. The boreal forest encircles the globe almost continuously just to the south of the Arctic tree line, and the same species found in our forest can also be found in eastern Siberia and western Eurasia.

The word "boreal" means northern. The best example of boreal forest in Nova Scotia occurs on the highland plateau of Cape Breton, where the dominant tree species is balsam fir. In other places, boreal forest is found near the seashore and in poorly drained areas where the predominant species is the black spruce. The extensive stands of white spruce seen on flat lands and slopes at lower elevations are not proper boreal forest.

White spruce comes up on abandoned farm fields and forms dense, even-age, easy-to harvest softwood forests, the favourite of clear cutters. Having worked in the woods myself, I don't know how else one can harvest these stands except by clearcutting. If only some of the trees are cut, the rest blow down. It is just about impossible to walk through a stand of white spruce, with its eye-jabbing dead twigs, and there is almost no herbaceous vegetation in the dense shade underneath.

Following a clear cut, the land comes up in more interesting, ecologically-diverse stuff like raspberry, birch, pin cherry and poplar. One day it may be a hardwood forest with wildflowers. Ralph S. Johnson agrees that clearcutting is the only way to harvest white spruce, but he is a forester and recommends immediate replanting with red or black spruce.

Regardless of how they are harvested, the big stands of white spruce across the province tell a story. That is the forgotten one of marginal farmland cleared, cultivated, farmed out and abandoned by bankrupted homesteaders failing to make a go of it.

A hundred or more years ago there was vastly more cleared land in Nova Scotia than there is today. I read that in 1900, 75% of the land in this province was cleared, and 25% forested. Today those figures are almost exactly reversed. Much land, formerly cleared, most of it not fit for cultivation anyway, has grown up in spruce woods to feed the pulpwood industry. Homesteads and settlements of those who toiled to work the land are abandoned and gone, marked only by caved-in foundations and piles of mossy stones deep in the forest.

The true boreal forest occupies land that could never have been cultivated. Black spruce and larch are tough customers that are willing to live in bogs and swamps, plan to live forever, and don't care how they look. In black spruce stands where trees are marginally big enough to warrant cutting, they look their worst. These trees find just enough nutrients to grow tall and spindly, bare except for a knob of green branches on the very top. Scanty tree growth allows sunlight to penetrate and permits understory.

Figure 98: Shinleaf pyrola

Vegetation leans toward mosses, sedges and ferns, as well as ericaceous shrubs. Bluebead lily, bunchberry, and twinflower are probably there, and undoubtedly wild lily-of-the-valley. On drier sites there will probably be wintergreens and maybe mayflower. Lady's-slipper is a good bet, and possibly other orchids.

Some of the most curious plants in the Flora are found under the spruces, including pyrolas, which are saprophytic, meaning they live on decaying roots and wood, and the ghostly white Indian pipe, which is parasitic on tree roots and has no green chlorophyll at all.

In treed bogs near the coast, nutrients are even harder to come by than in the swamp, but the spruce look happier. No one has the slightest desire to wade into the bog and cut them down. They may be a hundred years old, but hardly over six feet high and shaped like perfect little Christmas trees. In fall, their tops are loaded with round purple cones that a grown man could reach from the ground. Thousands of seeds are scattered, of which one or two may find a spot to grow.

Other trees might share the bog. Birches, red maple, mountain ash, and feathery-looking larches are common, themselves dwarfed down like the spruce.

Figure 99: Rhodora

This is really a sphagnum bog with trees in it, and the home of all the classic bog plants. Choice spring blooming shrubs such as bog laurel, rhodora, chokeberry and Labrador tea would be there. Pitcher plant and cinnamon fern are almost a sure thing, as are cotton grass, orchids, and probably the little sundew.

The forests of the boreal element are one extreme of the forest environment in Nova Scotia; the Alleghenian element is the other. The Alleghenian element is primarily a rich deciduous forest of hardwood species extending into Nova Scotia from New England and farther south. In this province, it is, or was, found only on the deepest and most fertile soils and was, for this reason, the first forest cleared for agriculture.

It is characteristic of the sugar maple woods and the river inter-

Figure 100: Wabanaki-Acadian forest

vales, and reaches its best development along the Cobequid Mountains east to Truro, and the hardwood hills and river valleys of Colchester, Pictou and Antigonish counties and Cape Breton.

The rich forests comprising the Alleghenian element were never widespread in Nova Scotia, and survive today only in scattered remnants, but still shelter some of the rarest plants in the province. Some of these are the spring ephemerals of the rich river intervales. Others, found in upland sugar maple woods might include Solomon's seal, bellwort, spring beauty, foamflower, red trillium, and round-lobed Hepatica.

Between the coniferous forests of the boreal element and the deciduous forests of the Alleghenian element lie the forests of the Canadian element, which covers everything in between. In general, these forests are "mixed" forest—a mixture of coniferous and deciduous species. This mixture can vary widely in proportion, with the nature of the resulting forest varying as well.

Roland's boreal element is still the boreal forest, but the Alleghenian and Canadian elements are combined under the umbrella of what is somewhat presumptuously called the Acadian forest. This is more properly referred to today as the Wabanaki-Acadian forest, as it is unceded land belonging to the original inhabitants.

A description of this forest comes from Jamie Simpson's book *Restoring the Acadian Forest*:

> The Acadian Forest covers much of the Maritime Provinces, and extends into New England and Quebec's Gaspé Peninsula. It is an area of transition between two larger forest ecosystems, the Northern Hardwood Forest to the south and the Boreal Forest to the north. As such, the Acadian Forest combines elements from each of these forests, creating a blend of softwood and hardwood trees (32 species in all) found nowhere else. In its natural state, the Acadian Forest is one of the most richly diverse temperate forests in the world.

So, let's get into this most richly diverse temperate forest and do some plant hunting.

Just about all the spots Mary and I explored in the Spring, the floodplains, the gypsum, the oak woods with its mayflowers, and the Cape Breton uplands, each a mixture of conifers and hardwoods, would fall under the umbrella of the Wabanaki-Acadian forest. Now, in mid-Summer, nature is moving on. The flowers we saw in Spring have faded, berries are ripening, shade is deep, and a new community of plants has taken over. We can look at these places all over again.

In the shade of the forest, things have changed. Shade-loving plants have big leaves. The ground is now so hidden, and the summer plants so tall, that it is hard to see where we are going or where to step. Revisiting places we were at in the Spring, when the trees were leafless and the forest floor wide open, we don't recognize anything. We have difficulty finding landmarks that looked so obvious earlier. Exploring the forest after June is hard to do except on a trail.

Fortunately, there are good ones.

Forest trails we have enjoyed in Summer include old woods roads, game trails, trails built by friends, and trails established by

Figure 101: Summer ostrich fern

parks people or volunteers in rural communities. A good summer trail is nicest near the water.

Some of our favourite trails lead to waterfalls, some follow streams, some are near a lake or the seashore. The cooling splash of a brook, the thunder of a falls, or the fresh breeze off the ocean or the lake makes all the difference. Plants are more beautiful near the water, and there is always the chance of finding a rare fern or a flowering plant we have never seen before. The water's edge affords a place to eat our lunch and dabble our feet to cool off until we are ready to go again.

When we go back to the intervale we visited in April, we follow

Figure 102: Meadow rue

the river. If the water isn't too low, we might even fish. The leaves of bloodroot and trillium are big and lie low to the ground, and their seed pods are swelling. We wouldn't see them if we weren't looking, because the entire intervale is now a forest of waist-high ostrich fern, cow parsnip and meadow rue, with orange-flowered touch-me-not thrown in.

The cow parsnip is hard to miss, with its great palm shaped leaves and wide flat heads of white flowers. The meadow rue is just as tall as the cow parsnip and ferns, but very delicate-looking. When it is first coming up, it looks like, and has been mistaken for, the maidenhair fern. With warm weather, it shoots up and blooms with loose heads of tiny white flowers.

Figure 103: Cow parsnip

In a sheltered spot near a bend in the river, at the foot of a big rotted stump where the Dutchman's bree-ches used to be, we find more rarities. Here are Solomon's-seal, Jack in the pulpit, and a small patch of dwarf ginseng.

The Solomon's-seal, which in June would have had pairs of bell-shaped flowers dangling from an arching stalk, now has ripening berries. Jack in the pulpit, with its three part leaves and striped hood, is in its prime and there are tiny ones coming up all around with just one leaf each.

Dwarf ginseng doesn't look at all like the ginseng sold in Oriental groceries. It is only a couple of inches high and has ball-

Figure 104: Jack in the pulpit

shaped clusters of small white flowers, followed by berries. It does grow from a small bulb which, if it were more plentiful, might be harvested for its medicinal properties.

The presence of these three species, as well as the bloodroot and nodding trillium, suggest that this particular floodplain is as rich as any in the province, and belongs in Roland's Alleghenian element. To prove it, we find a tall specimen of Canada lily blooming along the river bank.

Another forest we visit in summer is in Middle River, close to Baddeck, Cape Breton. We can stay with our friend Nancy, who guides us on trails she and trail builder Clarence Barrett have built along the Mill Brook, a tributary of Middle River and as beautiful a plunging, foaming stream as any in Cape Breton.

Today, August 15th, Mary and I are trying to keep up with long-legged Nancy, who hikes these trails every day and swims in the brook. She doesn't mind slowing down for us, though, so we can look at plants, because she is interested in them, too. I admit that I am looking for wild orchids, and pretty soon, I find a couple.

I am apprehensive because Nancy has brought along her dog,

and the dog is racing back and forth up and down the slopes as dogs do. I am worried that if I show any interest in a plant, the dog will step on it.

I know I should be fond of dogs, but I have never been comfortable taking them into the woods with me. They tend to tear up plants and range so far ahead that they scare away any wildlife before I see it. Of course, if we still had a dog, Mary would have brought it along and I would have had to be a good sport.

Today Nancy is watching her dog so I can relax. I no sooner tell Nancy that I am looking for orchids this summer, than I find one. Down in a damp sinkhole is a tall, thin, green orchid of some sort. I skid down the side of the sinkhole to look at it and nearly rub it out myself.

I am completely unprepared, which is unusual. I don't have the orchid field guide with me, or any other plant book. I don't have a camera. Even Mary doesn't have her cell phone to take a picture with. Sometimes I carry a plastic bag to cut off a branch and take it home to identify, but this is a single plant with a single stem so I would never cut it anyway. I try to look at it carefully and memorize what it looks like for later, but this never works.

Figure 105: Broad-leaved Helleborine

Sure enough, when I eventually consult Carl Munden's guide back at the car, I can't find anything that looks quite like what I remember. Or maybe several look like what I remember and I can't decide which one. I finally decide it must be tall leafy green orchid, but I will never know for sure.

Down by the stream, I find my second orchid. There are many specimens of different sizes. Some are in bloom, and some aren't.

I know this orchid. It has been following me around. It is what Carl Munden calls the "weed" orchid, the broad-leaved Helleborine.

The first time I "discovered" it, I was as excited as if I had found it in unspoiled wilderness. In fact, it was in my sister-in-law Janice's flower bed.

The second occasion was less exciting, and the setting even worse. This time we had rented a cabin in Maine for a week and the orchid was coming up in a sickly little flower bed of scrawny petunias. On the other side of the road, though, were some good woods and there were a lot of them there.

The novelty was wearing off. In the end I saw so many of these orchids in Five Islands Provincial Park that I quit looking.

These are rich woods along Mill Brook. Velvety mosses cover fallen logs,

Figure 106: Rose twisted-stalk

boulders and the ground itself. Great blocks of bluebead lily and bunchberry grow in the moss, along with arching branches of false Solomon's seal and twisted stalk, with its fruits hanging like cherries.

On either side of the trail are deep, dark and mysterious sink holes full of broken trees, ferns and, undoubtedly, botanical treasures. If I was younger and had more time I would slide down for a look. Along the stream, spray wets the rocks; and ferns, the most outstanding feature of these woods, thrive.

It is believed that the uplands of Cape Breton were only lightly touched by the last prehistoric glaciers that scoured the rest of Nova Scotia. Many species exterminated over most of the province survive in Cape Breton. This is particularly true of the ferns.

Every variety of wood fern is here, as is the glossy Christmas fern and the rock polypody that covers mossy boulders.

Along the brook, we find the cinnamon, the sensitive, and the

Figure 107: Braun's holly fern

royal fern. New York and lady fern are common. These are ferns of any rich woods in Nova Scotia, but two others, the male fern and the Braun's holly fern, we have only seen here.

Fronds of the male fern are leathery and smooth; those of the Braun's holly are furry with cinnamon coloured hairs. Both grow in imposing clumps up to four feet high.

Along with the mosses and other shade lovers, and the foaming and splashing stream, and Nancy and her dog, we've had a good day. Mary and I explored in other rich woods last summer, along good trails in the national and provincial parks and trails leading to waterfalls, and I will talk about these in another chapter.

Rich forests are shady and cool on a hot day, and boast their own distinct flora, but there are other sorts of woods under the Acadian forest banner. Mary and I seem to be attracted to the stunted kind, the ones at the other end of the fertility spectrum, growing in sand near the seashore. One of these is the oak forest and heath at the Pomquet Beach Provincial Park, and the other is what we call the "elfin" forest of white birch near our cabin.

This forest looks every bit elfin, as if it might be enchanted. It can be seen from a distance, a good-sized grove of low and crooked white birch, like a living bonsai. The birches cover the back side of a long sandy slope, which must once have been a dune.

It is a mystery how a little forest survives here, on abandoned land that long ago was pasture. The land all around is grown up in a handsome, shrubby sward of bayberry, winterberry, chokeberry,

Figure 108: Slender lady's-tresses orchids

alders and wild roses, with patches of low juniper and spruce. Open remnants of old pasture stretch up to the edge of the cliff where cranberries grow, and big chunks are caving off into the ocean.

There is no real trail up to the birches. From our cabin we walk along the shore until we come to the foot of the bluff. From there we head up on what used to be an ATV trail but is now more of a narrow game trail, where we must be careful not to be scratched by roses or tripped by blackberry vines.

At the top, where cranberries and three-leaved cinquefoil grow in the thin grass and I sometimes find the slender ladies'-tresses orchid, we take a right turn and head down through the old pasture to the birches. At the top of the hill where it descends into the birches are some old stones that look like they might once have been part of a foundation for some building. This is not surprising, because we know that fifty years ago most of this land was farms and pasture.

The birches descend the slope and then level out for a ways. They are old, probably more than fifty years, low and distorted. Still, their trunks are clean and white and for the most part they are nicely spaced and easy to walk through. There is no path, though, so we wander slowly at random to see what we can find. Birches don't cast as much shade as towering maples, and underneath is warm with patches of sun.

Figure 109: Cinnamon fern spore fronds rising

What we find right off the bat is all our favourite summer wildflowers and ferns. The ground tends to be damp and there are sizable patches of cinnamon, interrupted, and sensitive fern. Cinnamon and interrupted look much alike, and it is hard to tell the difference unless you can find the slender, rusty-coloured fertile fronds of the cinnamon fern, or a frond of interrupted fern with a congested section of spores "interrupting" the symmetry of the frond. We see that earlier there was blue flag iris blooming with the ferns, and

158

cattails grow some places.

In drier sections, the woodferns and the little oak and beech ferns mix with bunchberry, starflower, wild sarsaparilla and wild lily-of-the-valley. As in every other forest we have visited this summer, there is a great deal of Clintonia, or bluebead lily, with jade-blue beads ripening. In the sunniest spots, clumps of hawk weeds bloom bright yellow.

Figure 110: Interrupted fern frond with spores

Nothing uncommon here today, just the usual beautiful stuff.

We are finished exploring for the day, so we veer off towards the fox den to see if anyone is home. We do this each time we come. It still looks kept up and used but no sign of the fox. I guess he would have to be deaf not to hear us coming.

The den is dug deeply into the sandbank. The hill of sand taken out is compacted and smooth. The only time we have actually seen the fox was a few years ago when we saw him racing away and were tipped off as to the presence of the den. We are always careful to tread carefully and not disturb anything.

At this point, a game trail, which is undoubtedly the one the fox uses, takes us back up through the birches to the pasture, from where we proceed back to the cabin for a drink and a rest.

Finally, to wrap up summer in the forest, we visit our other favourite stunted forest, this time of oaks, in Pomquet. Today, August 10th, at the Provincial Park, it is a toss up whether to go to the beach, which is a particularly good one, or walk the boardwalk through the "slacks". I am with my daughter Margaret, and we decide to do both—she goes to the beach and I go to the boardwalk, that is.

The boardwalks take me through a very unusual type of Acadian forest, established on ancient sand dunes. White birch and red maple are present, but the most dominant trees are the red oaks, with their big, glossy leaves.

Oaks are quite rare in Antigonish County, but here they grow in almost pure stands. Due to the particularly lean soil conditions, mature trees are low, spreading, and older than they look.

Here and there are white pines and larch. The seasonal fall of oak leaves and evergreen needles cover and mix with the sand, creating perfect conditions for a very unusual understory—the heath.

Figure 111: Pomquet Heath, early spring

The heath is evergreen and nearly eternal. The word derives from the fabled "bloomin' heather" of the British Isles where expanses of colourful heather carpeted the moors. It is used to denote wild areas dominated by members of the blueberry family. That is certainly appropriate here.

Blueberry and bearberry are the dominant plants here, along with huckleberry and wintergreen. There is a great deal of Labrador tea, bayberry and low blue-green juniper. Lady's slippers have disappeared until next year. Mayflower plants are abundant but hidden under blue bead lily and patches of bunchberry. Cinnamon fern and other ferns are plentiful. Low shrubs that were leafless and unremarkable in the Spring have leafed out and bloomed.

Unlike the frenzy of exuberant growth that takes place in fertile forests, heath plants stick to business. This is the business of flowering and ripening berries. The most conspicuous plants of the heath are evergreen and slow growing. Plant cycles are stable year after year and the heath stays just about the same.

As we will see in the next chapter, heath is almost universal on Nova Scotia's seashores and barrens. It is as lush and low and as beautiful as anything in Europe, and high in plant diversity, but it is new-world heath. No bloomin' heather here.

Figure 112: Rest stop, Mica Hill barrens

8: The Barrens

Figure 113: The Barrens

If you think things have been interesting so far, just wait until you visit the barrens. Here is where you glimpse eternity. Everything was in place in the barrens by the time the glaciers began to melt 14,000 years ago, and not much has changed since. The mighty sheets of ice that scoured the Precambrian rocks into gravel, moved boulders, carved out gorges and pushed up drumlins, eskers, escarpments and islands, dropped everything and beat a hasty retreat, if you can call a change taking 4,000 years 'hasty'.

Scientists believe that it was 10,000 years ago when nomadic hunters wandered into what is now Nova Scotia, following the re-treating glaciers and the caribou. Large herds of caribou called the barrens home, for the barrens are where the reindeer moss, *Cladonia*, grows in abundance.

Their numbers supported bears, wolves, and other carnivores.

163

Caribou fed and clothed bands of nomadic hunters and their descendants, the Mi'kmaq of today, for 10,000 years. White men arrived and the caribou were gone in 200. An all too common story of colonization and greed.

The modern human being, following his extermination of a fellow creature, especially a big, gentle one like the caribou, is invari-

Figure 114: Reindeer moss

ably stricken by remorse and nostalgia, and resolves to bring them back and make it all right again. This never works. There are too many things we don't understand.

The caribou that roamed Nova Scotia were known as the woodland caribou. In the summertime they roamed the open barrens and bogs, and in winter moved to the shelter of the deep forest. They subsisted mostly on lichen—the reindeer moss in the summer and the old man's beard and other woodland lichens in the winter.

All was well until European settlers began cutting down the forest and hunting. In the winter, caribou depended on the shelter of mature forest. With lumbering and clearing for agriculture, mature forest began to disappear.

The long, narrow shape of Nova Scotia enabled hunters to predict the migrations of the caribou. Hunters shot them in large numbers to feed the men working in mines and lumber camps.

As the old forest was cut down, and replaced by succulent regeneration, the whitetail deer moved in. This was the final blow. The deer lived with a brain parasite that didn't hurt them but was lethal to caribou. The last caribou in mainland Nova Scotia was seen in 1905, and the last in Cape Breton in about 1912.

Within a few decades, hunters and nature lovers were wishing they had the caribou back. Abundant reindeer moss was still there on the barrens and there seemed no reason that caribou, in the absence of hunting, couldn't thrive again. In light of that, 51 caribou from Quebec were released into the Cape Breton Highlands National Park in 1968 and 1969.

It didn't work.

The animals weren't hunted, but there wasn't suitable forest for them to winter in. The brain worm spread by the deer was deadly. The last of the re-introduced herd was seen in 1972, and the experiment has never been tried again.

Incidentally, wildlife wardens, anxious to get some big deer established in the Cape Breton highlands, released 18 moose from Alberta in 1947 and 1948, prior to the failed caribou project. It took time to find out, but this was a failure of another sort.

Fresh new tree growth following the spruce budworm infestation in the 1970s was just what moose were looking for. The population exploded, and today moose browsing young trees prevents the regeneration of boreal forest, turning large areas of the highlands into unproductive meadow. Those thousands of moose in Cape Breton are all descended from only 18 brought from Alberta.

Even more astounding is the fact that all the moose in Newfoundland, a province overrun with them, come from just four animals imported from New Brunswick in 1904. There, too, they are causing a lot of problems.

I guess it is clear by now that, to have a healthy wildlife population, the best idea is to not exterminate them in the first place.

Today, the barrens are another near-wilderness because there is nothing there anyone wants. Only the occasional fisher, berry picker or botanist ventures into this forgotten landscape, where the wind blows across the shrubby heath and the reindeer moss awaits the return of the caribou.

The barren habitat is a varied one, the common denominator being chiefly its apparent emptiness and austerity, It is defined in a N.S. government publication as "a rocky heathland with dwarf shrub and lichen vegetation that occurs in Nova Scotia along the Atlantic coast as well as inland."

'Rocky' is an understatement in many barrens, where rocks dropped by glaciers can be the size of houses or trucks.

Heathland itself is defined as "a vegetation type dominated by low growing shrubs from the heath family (*Ericaceae* also including *Empetraceae*)". This rather scientific definition tells us that the heath is comprised mostly of species from the blueberry and black crowberry families, which are very large families, and often evergreen. Plant succession is very slow if not absent on the barrens, and the heath is almost eternal.

Barrens may have arisen from other forces besides glaciation. A prerequisite, though, is scanty soil and an exposed location. Whatever removes or degrades the soil to the point that trees can't grow can result in barren formation. Repeated burning, often associated with deforestation, will do it. Many of the large barrens in southwestern Nova Scotia have arisen in this way. A hardpan or underlying bedrock can prevent tree roots from penetrating and result in barrens; and extreme exposure, which is almost a given where barrens occur, is a factor.

Barrens of all sorts are defined as heathlands, but the species composition of the heath in different barrens can vary considerably. In 2020, botanists from Saint Mary's University in Halifax published the results of a seven-year survey of barrens ecosystems in Nova Scotia. Caitlin Porter, Sean Basquill, and Jeremy Lundholm surveyed barrens sites from one end of the province to the other, and described twenty-two distinct barrens associations based upon topography and vegetation.

Their findings concluded that the major determining factors of plant composition in the heathlands were soil moisture and exposure to winds and salt spray. The twenty-two barrens associations included two herbaceous associations where shrubs were not an important component; fourteen dwarf shrubland associations with various combinations of low, sometimes prostrate woody shrubs; and six larger shrubland associations.

The Atlantic coastal barrens extend from Canso through Guysborough and Halifax counties at least to Peggy's Cove. Coastal barrens may abut boreal forest of spruce and fir on the landward side, then resume at higher elevations.

Probably the greatest area of barrens in the province occurs on the plateau of the Cape Breton highlands. The barrens that Mary and I visit are above Cole Harbour in Guysborough County.

It's not easy to find your way into these barrens. There are not many people living around there and no real roads into the interior. Local people probably know of ATV or Jeep roads, but the only road we know of is the one to the radar station. There's no timber or fish in these barrens; no coal, no gold, no agricultural potential, but there is lots of bare rock. Perfect for a radar station.

The radar station in Cole Harbour was built by the RCAF during World War II in the 1940s, to survey the skies for aircraft in distress, dangerous weather, and Germans. It was operated by a staff of up to 36 airmen at a time, and was located at the edge of the barrens on a gravel road between Cole Harbour and Queensport. By now, the station has fallen into ruins, leaving only concrete slabs and abutments, and massive chunks of rusting iron.

On the paved #316 highway at Cole Harbour there is a parking spot and a plaque to commemorate the former radar station. From the parking lot, the old road to Queensport begins, and passes by the radar station site on its way.

The road is very badly eroded at the start. We have seen ATVs using it, but even they have trouble. It is about two kilometres to the old station, and we hike it on foot.

The first kilometre rises steeply and, because we are botanists, we are not in a hurry. The next kilometre is easier and leads to the site, which is now identified by a gazebo with more historical information.

Today, November 17th, I am alone. Mary is working. The day is quite sunny and warm for the season, and before I even get started I hear, then see, a sizable waterfall very close to the car park. We have only been here before in the summer, and never noticed the waterfall. Maybe it is seasonal. I try to get close because I never pass up a chance to check out a waterfall.

Unfortunately I can't reach it. Hurricane Fiona has knocked down so many trees below the falls that I can't get past.

The road is clear, though, so I head up. This road is old and looks old. There has probably not been any grading or improvement since 1945.

That explains the abundance of mayflowers, wintergreens (two kinds), bearberry, blueberry, bunchberry and twinflower on either shoulder. As a matter of fact, I see more mayflower here than I have seen anywhere else. Of course, today, on the doorstep of winter, none of these plants are blooming; but they are evergreen and can be identified by their leaves. Come spring, the show should be spectacular.

Before long I make it to the gazebo. The area around is grown up with chokecherry and alder, and little paths lead through to what is left of parts of the old radar station.

Since I have been here before, I skip all that and take the path that leads to the barrens. Just before the steps up to the gazebo, to the left, there is an ATV trail. It passes through the bushes, then a grove or two of trees and then, suddenly into the barrens with a view for miles in every direction.

To the South is a small lake which is probably the source of the waterfall. Beyond that is the ocean. To the North and West is a jumble of giant boulders and unbroken heath. To the East is more of the same, plus a sparkling lake bigger than the first one.d

The two lakes have formed in hollows scooped out by the same glacier that dropped off the boulders a mere 14,000 years ago. It looks as if it might have taken place recently and that maybe the glacier is hiding on the other side of the hill.

The ATV tracks continue another half kilometre or so to a sort of spire or flagpole that is embedded in concrete. There is a picnic table, which is a good place to sit and observe the lake.

Today, the wind is too cold and brisk for sitting, so I keep on the track that leads on into the distance. All around are lichen-covered boulders and the heath. Here and there are mounds of spreading juniper or a spruce or larch flattened by the wind. Birches too, are low and flattened. All a testament to the savage cold and winds of winter.

Entering the barrens is as if walking into a vast, life-sized Japanese bonsai garden. Certain low, woody shrubs and herbaceous plants are tolerant of the thin soil and winds of the barrens, and constitute the dense heath, with giant boulders and wind-twisted spruce or birch scattered at random. Lakes are perfectly placed to set it all off. A massive chef d'oeuvre conceived and planted by the Master.

The muddy ATV trail continues incongruously in a straight line through the heath into the distance. It is an easy path to follow, and I take little trips off to the side when I want to look at things.

Walking across the barrens without a trail is not easy. The heath appears to be flat, and can indeed be walked upon with care, but it is bouncy and deeper than it looks. It is easier right now than it is in the summer, because the shrubs have lost their leaves and I can see the ground. Numerous flat outcroppings of rock rise a bit above the heath and I make from one of those to another.

I am only looking at vegetation, so I don't have to go far. I could, though, if I wanted to. The barren is vast. Maybe I could discover a lake to name after myself.

Truth is, I am afraid of losing my trail. I haven't gone very far yet and, if I had to, I could find my way back across the heathland without a trail, but it wouldn't be easy going. So I stick to the trail and follow it until it dips out of sight over a hill. I am curious to see where it is going so I go over the hill myself, only to see it continue on in a straight line as far as I can see.

By now, I have lost sight of the lakes at the start of the trail, I don't recognize any particular boulders, and the heath looks the same in every direction. I realize that if I didn't have the trail to lead me back, I could get seriously turned around. Without a compass, that would mean lost.

I have heard stories of people lost in the barrens, and I wouldn't want to be the subject of another one. This is only idle musing, because I do have a compass in my back pack and the trail to follow back, but I think if a person wandered into the barrens unprepared, getting lost would be very easy.

The Bonnet Lake barrens of Guysborough County, which is where I am now, is shrubland association S2—Dwarf Huckleberry

Figure 115: Bonnet Lake barrens

Heath in the survey. Huckleberry is the dominant species, and to go with it the list includes companions such as bayberry, Labrador tea, sheep laurel, Rhodora, black chokeberry, blueberry, and black crowberry.

Virtually every one blooms beautifully in the Spring, many with pink or white bells characteristic of plants in the blueberry family, and they turn brilliant shades of yellow, red and orange in the Fall. Most produce berries.

The first time I visited this place, blueberries were still hanging on the plants as their changing leaves coloured the entire barrens red. Today, in mid-November, leaves have fallen in preparation for winter, and the barrens are a uniform but pleasing sward of grey bushes and boulders, with occasional creeping juniper and stunted spruce affording splashes of green.

Figure 116: Mosses and lichens

The variety of herbaceous plants in the heath is more restricted, especially at this time of year after the leaves are gone. I find some pitcher plant, and reindeer moss lichen flourishes with no caribou to eat it. The survey says there would be cinnamon fern, golden-rod, gold thread and wild lily-of-the-valley, if I had been here earlier.

The road back to the car descends into the shelter of the trees, and the difference is like night and day. Plants here, protected from the wind and cold, are green and growing and, unlike those in the lifeless barrens just up up the road, seem unaware of approaching winter.

I promise myself to come back with Mary in spring when the mayflowers along the road, and most everything in the barrens will be in bloom.

May 23, 2023. Back to the radar station and barrens to check on all those mayflowers and see what is growing in the barrens. We packed our lunch and left early in the morning, driving down the 316 highway from our place to Cole Harbour.

Pulling in to the parking at the radar station monument, we saw trouble. There was an excavator parked there, and what we could see of the nice old trail to the barrens was all chewed up. Trees had been cut down and the road scraped to the width of a two-lane highway. There were mounds of gravel waiting to be spread, There was nothing left of the wildflowers. I almost cried.

Granted, this lower part of the road had been in particularly bad shape and almost impassible even for ATVs. The old radar station site is beginning to be promoted as a tourist destination, though, and we feared that this road "improvement" was meant to go all the way up.

Nevertheless, we started slogging our way up through the loose gravel and devastation. To our immense relief the roadwork left off after half a kilometre or so, and for the rest of the way it was back to the sweet little road with two wheel tracks and mayflowers in the middle.

We hit the mayflowers right on. We had never seen such masses in any other place. In spots, they were almost a ground cover. Typically, only a small percentage were actually flowering, but considering the number of plants, that was a lot of bloom.

The mayflower companions, the wintergreens, bearberry and foxberry, bunchberry and wild lily of the valley combined to cover the roadside and would soon be coming into bloom.

From the radar station we made our way into the barrens. The day was warm and there was a good breeze blowing to keep off the flies. The empty barrens with its boulders and little lakes was as invigorating as ever.

The plants in the barrens were budding, but not so far advanced as those along the road. The dwarf huckleberry, the signature plant of these barrens, looked like it might be in bloom in maybe two weeks. Blueberries were beginning to bloom; so was bog rosemary, and Labrador tea wouldn't be far off.

On this day in May, despite the islands of dwarfed spruce and juniper, the barrens still looked rather grey. Except for the mosses and lichens, that is. These were alive and spectacular. The reindeer moss that grew everywhere was soft, deep and creamy white.

Buried in it were the reddish pitcher plants just beginning to thrust up. Sphagnum mosses of all kinds were springing up with fresh new growth in every shade of yellow, red, orange and green. Patches of colourful lichens covered the boulders. In two more weeks the shrubs should all be in bloom.

It is becoming obvious that, as with every other environment we visit, we will have to visit the barrens time and again to see everything.

We returned home, hoping that the excavator was finished and only fixing up the lower part of the road. We won't be back for a

Figure 117: Mica Hill barrens

while to find out, though. We have another barrens to visit.

July 1, Canada Day, we camped in the Cape Breton Highlands Park with the intention of hiking the Mica Hill trail, which leads up into the barrens of the highland plateau. These barrens are the most extensive in Nova Scotia, and we had never been up there.

The trail is 7.9 kilometres round trip, and described as "gradual climb through Acadian and boreal forest to taiga barrens, panoramic view of highlands plateau." It is all that and more.

The Acadian forest segment is pristine and intact. In its shade it shelters handsome patches of the woodland wildflowers and ferns. Little rivulets of water trickle down the mossy bank here and there.

The trail climbs slowly into the conifers of the boreal forest, and then out of the trees into the barrens. The view is breathtaking.

Figure 118: Bunchberry climbs a stump in the barrens

This is a dwarf huckleberry barren so there is a lot of two-foot-high huckleberry shrubs just now blooming with little red flowers. Mixed with the huckleberry are great areas of spruce, fir and larch, pruned and dwarfed by the wind. Most are only a foot or two high but very old. It looks like a huge planting of dwarf evergreens at some botanical gardens.

Around and beneath the huckleberry and evergreens is a thick carpet of reindeer moss, through which grow and twine the herbaceous plants of the barrens. Wild lily of the valley and bluebead lily are common, and I guess we will see wild sarsaparilla and bunchberry wherever we go. Nothing stops them.

In damp hollows, the dark red flowers of pitcher plant rise above the moss. Scattered throughout are button-like flower heads of Labrador tea, while creeping types of blueberry and cranberry

Figure 119: Pitcher plant and Labrador tea

nearly cover the ground.

We discovered a new sort of blueberry, which is a tiny, woody shrub only three or four inches high. It bears the interesting name of alpine whortleberry, or bog bilberry. Take your pick. It has a good berry, although it didn't look like there would be many of them.

There was a great deal of Rhodora that had mostly finished blooming and just as much pink sheep laurel just beginning.

The trail is a long one, but easy to manage, and twists and turns through little valleys and hills. At strategic spots, the parks staff have placed benches to sit and admire the view. We stopped at the very first one and were so impressed that we practically used up our camera batteries snapping pictures of taiga, with towering blue mountains and Aspy Bay in the distance.

The farther we went, the views only got better and better. Sometimes there were chains of lakes far below. We could point our cameras in any direction and get a wonderful shot. We laughed about taking so many photos at the very first stop.

We finally put the cameras away and concentrated on hiking and looking at plants.

We figured that from the start of the trail to Mica Hill should be about four kilometres, since the round trip is about eight. When you are poking along like we do, checking out the plants, a kilometre is a long way. Every time the trail descended into a valley, then began to climb another hill, we figured we must be there. If this isn't Mica Hill, then the next one had to be, but none of them were.

We were getting tired by this time and the trail went on and on.

Finally we came to a marker that said 'three kilometres'. Criminy, another kilometre to go. We stopped to rest.

In the distance, the biggest hill of all rose above us like Everest. Its top was covered in gleaming white stone like snow. Well, that was Mica Hill for sure. The trail description said that it was capped with mica and snow-

Figure 120: White lady's slipper along the trail

white quartz. At that point we began to call it Mica Mountain. It was no hill.

After we had rested, we took up the trail again. It wasn't as bad as it looked.

Along the way we were entertained by the sight of many lady's slipper orchids, both pink and white, still in bloom. At lower elevations they were finished long ago.

Climbing in elevation turns back the season, like turning back the clock. Plants that have finished blooming at lower elevations may not even have started at higher ones. I was told that going up 1000 feet in elevation is the same as going 400 miles further north.

My brother lives at 8500 feet above sea level, in the mountains of New Mexico. When you work it out, he lives farther north than I do.

Figure 121: Mary on Mica Hill

Mica Hill, or mountain, was neither as high or as far away as it had looked when we were tired. Reaching the top was very invigorating, for there was a cool, brisk breeze and the snow-white boulders really gave the impression of snow.

Between and around the white boulders were sheets of mica, up to two or three inches long, that reflected the sun like mirrors. The reindeer moss, the little alpine whortleberry, crowberry, foxberry and three-leafed cinquefoil filled in the gaps.

For an hour or so we were the only people at the top, and had seen only one or two people on the trail. Then a couple visiting from France arrived. We saw a few more hikers on the way down, but, considering it was Canada Day weekend, this particular trail doesn't seem very heavily travelled. We would certainly recommend it to anyone who wants to get a taste of the barrens.

Mary and I have visited only a few of the twenty-two different types of barrens across the province but I will describe one more.

This would be classified a black crowberry barren, and is found at a place called High Head, off the Prospect Road near Halifax. As

Figure 122: Black crowberry barren

opposed to the other barrens we have described, this barrens begins on the rocks directly above the ocean and stretches off into the distance.

The black crowberry is a creeping evergreen with tiny leaves resembling the needles of evergreens. It covers the ground deeply, including all but the largest boulders. Mixed with patches of blue-green juniper, it is like a thick cushion to walk on, much softer than the prickly huckleberry and taiga of the other barrens. Here and there outcrops of boulders and low evergreen trees rise up, and in the distance are groves of thick, dark spruce.

It was Sunday May 14, when Mary and I and our daughters went hiking there with a friend. It was quite early spring, and the crowberry was not blooming yet. Here and there, twigs of Canada honeysuckle jutted up through the crowberry cushion, blooming prettily with their twin yellow bells. Most everything else we found, the wintergreen, blueberry, cranberry, and wild lily of the valley had yet to bloom.

In one spot on the trail was a mass of tiny white flowers which had us puzzled until we recognized the small leaves of goldthread. In another spot was a patch of wild strawberry in bloom.

Along the trail on our way back to the car, the serviceberry bushes were flowering.

So that is it for our explorations in the barrens. We hear the wa-terfalls calling.

Figure 123: Myles Doyle Falls, Cape Breton

9: Dripping cliffs, waterfalls and ferns

Figure 124: A hidden waterfall

A summer excursion you won't forget is one that takes you to a waterfall, and Nova Scotia has plenty. Mary and I try to hike to as many as possible of those we read or are told about.

Trails to waterfalls are almost always good ones for plant hunting, because they border the stream. Then there is the excitement of reaching the falls, where you swim or fish, laze as you eat your lunch, and poke around to see what is growing.

The waterfall, in its plunge, sprays and wets the crevices, grottoes, boulders and vertical cliffs alongside. Falls and "dripping cliffs" come up time and again in the Flora as the chosen environment of rare flowers, and ferns.

Botanists searching for Arctic and alpine disjunct species stranded in the uplands of Cape Breton focused on the cliffs and falls in the deep river gorges. In their study of five river gorges, published in 1969, R. W. Hounsell and E. C. Smith conclude:

> A common denominator for the three most productive stations of this study is a cool habitat. In each case trickling and seeping water was present in varying amounts, moistening the shaded cliffs. The water and the shade are probably not ends in themselves, important though they may be; it is likely that their most important function is in contributing to the coolness of the habitat. The temperature-moderating effect of a river running by a cliff, in combination with shade and seepage, could well be a telling factor in determining the suitability of the habitat for Arctic-alpine species.

Figure 125: Dripping cliffs and ferns

Of course, rare Arctic-alpine species are not the only species found in the vicinity of waterfalls and dripping cliffs. The coolness, mist and humidity are a haven for many. Hounsell and Smith report finding the little harebell, or bluebell of Scotland, at all their sites.

They also found the pretty little wild primrose. Others listed were meadow rue, wild sarsaparilla, hawkweed, yarrow, and even dandelion, none of them rare, and poison ivy, which I hope soon will be.

Figure 126: Bluebells of Scotland

From my own experience, I would like to add coltsfoot to the list. We usually expect to find coltsfoot in the hard gravel beside the road, where it blooms in April—the very first flower of spring. It is not one to pass up a good thing, though, and we find it growing luxuriantly by the falls and alongside streams.

As might be expected, Hounsell and Smith observed a great variety of ferns; some of them rare, most of them not. Ferns and water are inextricably linked together. Some ferns, it is true, are found in rather dry locations, but all ferns need free water in order to reproduce.

Ferns are a special class of plants that reproduce by spores rather than seeds. Ferns appeared early in the evolutionary history of green plants and have changed very little over tens of thousands of years. Prehistoric fern fronds, almost identical to those of the present day, are found fossilized in stone.

The life cycle of the fern is very interesting and unusual, involving two entirely different types of plant. Fern spores are truly microscopic, a fine dark dust released from what are called fruit dots beneath the frond. Spores must land on a damp surface to germinate and grow. When a spore germinates, it doesn't become a baby fern right away.

Figure 127: Beech fern and New York fern on a cliff

The fern has two generations, one called the gametophyte generation and the other the sporophyte generation. They are vastly different. The sporophyte is the fern we are familiar with, and produces the spores. The gametophyte grows from the spore.

The gametophyte is much prettier than its name. The individual prothallus—I don't like that name either—that grows from a spore, is a tiny, translucent, emerald-green thing, like a bit of moss. When many grow together they form a carpet much like moss.

At one time Mary and I grew ferns from spores and sold them. I never tired of looking at a flat of fern prothalli through the magnifying glass, they were such a lovely shade of green and the light shone through them. We kept the flats damp and shaded, and misted them regularly with distilled water.

The prothalli in the flat produce egg cells and sperm-like gametes that swim in the film of water. Fertilization takes place and before long newborn baby ferns pop up in the tray.

When the babies were a couple of inches tall, we picked them out of the tray and potted them into bigger and bigger pots until they were ready to sell. The mature ferns might one day be four feet high, but they all got their start from prothalli in a tray.

In the real world, of course, prothalli don't grow in a tray. If you know what they look like, you will find them on rocks by the falls or sometimes on wet logs or patches of damp mud. Sometimes you will see tiny ferns beginning to grow from them.

Ferns found in drier spots in the woods or on dry, rocky slopes require small damp spots to germinate spores, where they will be wet by dew and rain.

The various species of ferns are sometimes identified by the shape of their fronds, but more often by the size, shape and position on the frond of their fruit dots, or sporangia. Fruit dots are most commonly found underneath the frond, but in some ferns, such as the interrupted fern, the Christmas fern, or the royal fern, they may be densely clustered in a part of it.

Some ferns have both sterile fronds and fertile fronds. The sterile fronds are the fern-like ones. The fertile fronds produce the spores and don't look like fern fronds at all.

The fertile frond of the cinnamon fern looks like a two-foot-high fuzzy, elongated, cinnamon-coloured Q-tip. The fertile fronds of ostrich and sensitive ferns look like sticks covered in dark beads.

I count thirty-eight species of ferns in the Flora, which is not unreasonable. I would like to find them all, but that will probably never happen. Some are very rare.

Some of the rare ones were seen around the falls and gorges of Cape Breton by Hounsell and Smith in their study. These include the slender cliff brake, the green and maidenhair spleenworts, the alpine and the smooth woodsias, and the fragrant fern, which I look for every time I am near a falls, but have never found. The bladder fern and the bulblet bladder fern were more common, but still restricted to wet rocks and dripping cliffs. The rusty woodsia, the Braun's holly fern, and the marginal wood fern were listed as occurring on dry ledges and talus slopes nearby.

Not all waterfall-loving ferns are rare. Hounsell and Smith may have missed them, but you should be able to find the rock poly-pody, the beech and oak ferns, the sensitive fern and what they call the flowering ferns, cinnamon and royal, if you visit enough water-falls.

Figure 128: Rock polypody fern

Now for some true life adventures.

I can remember, I think, most falls I have hiked to or seen. If there are any I don't remember, I can't remember which ones they are. How is that for logic?

Each waterfall experience is memorable. It might be the trail, the stream, or the falls itself, or any combination of the three. Here are some I certainly won't forget.

Hounsell and Smith surveyed five different waterfall sites in their search for Arctic-alpine disjuncts. One of them was the Indian River, which I knew had a beautiful falls somewhere upstream. I was at the Indian River with my son Austin a few years ago, head-ing up to the falls, but I wasn't looking for disjuncts. I was swim-ming upstream with my fishing rod in my teeth.

This was my second serious attempt to reach the falls. The first was by a trail that went from the highway bridge and followed the crest of the gorge above the river. Some boys I talked to by the bridge assured me that the trail went to the falls.

I was with a friend and we had forgotten to ask the boys how far it was to the falls, but we were good hikers and had all afternoon. To make a long story short, we walked for an hour or more without seeing or hearing the falls. The trail began to get harder and harder to follow, and then turned uphill away from the river. Finally it just petered out.

We returned back down to above the river and tried to keep on without a trail, but it was too hard. We couldn't hear a waterfall or even see the river anymore, so we turned around defeated and went back to the car.

At the highway bridge where we had parked, the river is deep and dark and it looks like the obvious way to get to the falls would be to rent a canoe and paddle it upstream. That was the plan when I went with my son. There was even a canoe rental place just on the other side of the highway.

We got there on September 15th, our mutual birthday, and a beautiful day after a dry summer. With one glance, we could see that canoeing was out. The water above the bridge was low—too low. We could wade across it without going over our knees.

It looked, in fact, like we could probably follow the shore all the way to the falls, or wade if we had to. So, anticipating some good fishing below the falls, we grabbed our fishing poles and a little knapsack with food and tackle, and set off.

Everything went according to plan at first. We walked along the shore, and waded a bit, and made good progress. But then, little by little, the river began to narrow and the banks to get steeper. Soon we couldn't walk along the shore anymore, and were wading, sometimes up to our chests.

A little farther and the river channel became a narrow passage between vertical walls of rock and the water would have been over our heads. We were beaten once more.

This was disappointing. We had come a long distance and should be close to the waterfall.

Maybe there was another way. If we couldn't wade, we could swim, but the water was swift and we had our fishing poles. We couldn't swim with just one hand.

Well, a beaver carrying a stick doesn't swim with one hand, either. He puts the stick in his mouth. That's what we did. The poles came apart in two pieces. We clamped them in our teeth and began to swim.

It wasn't easy, but here and there the stream flattened out a little bit again and we were able to touch bottom with our feet and wade.

We made it to the falls. I would like to know if Hounsell and Smith had to swim.

The river certainly wasn't low going over the falls. The pool beneath the falls was so wide, so deep and so turbulent that we couldn't get near the actual waterfall.

The day was still sunny and warm so we sat on the shore and dried out. Austin caught a nice trout on his first cast and I thought we would be going home with a full stringer, but for some reason that was the only one biting.

I never got close enough to the rocks beside the waterfall to look for fragrant fern.

Another sunny summer day, Mary and I and our daughters took to the trail to James River Falls, a substantial waterfall near where we live in Antigonish County. The trail is quite a long one, maybe three or four kilometres, and crosses back and forth over a little stream.

We were hiking at a good pace, and so not paying much attention to the little plants growing along the way. We should have been, though.

The trail starts in and passes through the hardwood forests of the Pictou-Antigonish highlands, where I am told there is an unusually-rich assortment of native wildflowers. They probably were done blooming, anyway, and we would do our plant hunting when we got to the falls.

The last part of the trail is a steep descent into the chasm of the James River below the falls. This wasn't the first time we had hiked to these falls, and each time is good all over again. The waterfall comes down in two steps into a large pool that is a good place for swimming. Rocks on the shore are well placed for a picnic and sitting in the sun.

After eating, and sitting in the sun for a while, I noticed the flowers and ferns at the foot of the falls, and decided to go have a look. Maybe this will be the day I find the fragrant fern.

Getting to the foot of the falls wasn't easy, over rounded, wet, slippery boulders and cobble. I struggled with difficulty to a copse of bushes beyond which I could see some interesting ferns.

I had no sooner started to push through the bushes than I was attacked by a swarm of hornets. I had unwittingly put my head into their nest and they resented it.

Only two choices—run back across the slippery rocks and break my ankles, or die where I was from hornet venom. I ran.

I shot across the rocks at full speed and dove into the water fully clothed, directly in front of my startled wife and daughters. I stayed underwater long enough to put the hornets off my trail, then surfaced and crawled out to inspect my ankles and put mud on my stings.

I can only say two good things about hornet stings. The first is that you never see them coming. The second is that they are soon forgotten.

We returned late in the day to the car. Everyone agreed that it had been a good hike and that we would do it again. I was left wondering if that may have been fragrant fern just on the other side of the hornet nest.

At the next waterfall we visited, I was sure I would find the fragrant fern. *The Flora of Nova Scotia* said it was there. The place was called Hartley's Waterfall and it is near Mulgrave, by the Canso Causeway.

By now, you must wonder why I am so eager to find the fragrant fern. It is just that I want to see it. The description in the Flora goes like this: "a small distinctive fern with fragrant glandular fronds." Enough to make a botanist's heart beat faster.

Also, in those days we were selling ferns, and I thought we

might try growing it.

I was all for heading for Hartley's Waterfall as soon as I learned that the fern had been found there. The trouble was, the waterfall wasn't on any map, and a little aimless searching around Mulgrave didn't get me anywhere.

Then Mary reminded me that we had a good contact in Mulgrave—a very good woman and a gardener. Lillian Williams. We did landscaping work around her house sometimes. She wasn't doing everything herself now that she was a hundred years old. She did know everyone around Mulgrave far and wide, and was certain to know where Hartley's Waterfall was.

Lillian came in once or twice every summer with her daughter to the garden centre where we worked.

Figure 129: Hartley's Waterfall

Her daughter was over seventy, both of them sharp as tacks.

I was at work sometime later when I spotted Lillian with her daughter. I struck up a conversation, which wasn't hard to do with avid gardeners, and I asked her if she could tell me how to get to the falls.

She knew immediately what I was asking, and said, "Go see Harald Marr in Part Harbour. He'll tell you where it is."

At least that sounded like what she said. I looked but couldn't find any Part Harbour on the map, and no Marr in the phone book.

That evening at home I repeated the story to Mary. She laughed and said, "That's just the way they talk in Mulgrave, Bruce. Lillian told you to go see Harold Meagher in Pirate Harbour."

Well, Pirate Harbour is right next to Mulgrave, and there was a Harold Meagher in the phone book. I phoned him and we met at his nice place by the strait. He served me tea and we chatted a while.

He was an old man and had lived in Pirate Harbour all his life. He had lots of stories to tell about the days before the Canso Causeway was built and Mulgrave was the busy mainland terminus of the ferry crossing the strait. Even the train, which ran regularly in those days, had to come off the tracks and cross on the ferry.

Finally I asked him if he could draw me a map or explain to me how to find the falls.

He said, "No, I'm going to take you there."

So he got in his truck and I followed in mine and we drove a piece up a gravel road to a place where there was just some ATV tracks heading down into the woods. No sign or anything. Not even a place to park the truck off the road.

He said. "This is it," waved and took off.

I never would have found it myself.

That afternoon I went down alone and discovered the waterfall. The stream was medium-sized and clear and cold. At the falls it dropped into a narrow runway walled with vertical slabs of stone. It was possible to see the speeding water by holding onto a tree and leaning over the edge, but it took courage.

Deep in the runway, the stone slabs made an abrupt right angle bend, which the speeding water slammed into, then straightened out into the pool at the bottom. In the fading light, I could see that these falls were a fern wonderland, with polypody all over the rocks and beech fern at the foot of the falls. There were other small ferns on the dripping rocks, but as far as I could tell, no fragrant fern.

I returned soon after with Mary. We marvelled at the beauty of the place, and descended with care to the foot of the falls. Again, no fragrant fern, but a splendid place for a picnic. We have been back several times since with our daughters and our friends.

The fragrant fern still eludes me. It was reported from Hartley's Falls in the 1969 edition of the Flora, but it's gone in the 1998 one. It snuck away.

I'll just keep looking.

Figure 130: Corney Brook Falls

On June 26 last summer, Mary and I drove to the Cape Breton Highlands National Park to hike to the falls on Corney Brook.

The Corney Brook gorge and falls are the very epicentre of rare plant discoveries in Cape Breton. Starting in 1980, Harold Hines spent a couple of summers in the gorge, and didn't come away empty handed. Among his discoveries were the butterwort, the yellow mountain saxifrage, the small flowered anemone, and the white flowered alpine, *Diapensia lapponica*, all rarities, and the even rarer Lapland Rosebay rhododendron.

Rhododendrons are plentiful in Nova Scotia, found in bogs, barrens and everywhere that the soil is damp and sour. These rhododendrons, though, are all Rhodora—all the same species. They put on a good show when they bloom in lavender or sometimes white in early May, and they turn bright colours in the Fall, but they lose their leaves.

The big evergreen rhododendron, the Great Laurel, which caused such a stir when it was discovered near Sheet Harbour around 1860, has vanished from the province. The Lapland Rosebay is no Great Laurel and doesn't stand six feet high, but it is evergreen, rare, and an exciting discovery. Just listen to the description in the Flora:

Lapland Rosebay is a dwarf species, forming aromatic mats up to 60 cm. across. It has roughened branches, with thick wrinkled leaves up to 2 cm. long. The corollas are royal purple, and the flowers form capsules 4-7 mm. in length.

Figure 131: Lapland Rosebay rhododendron

It has only been discovered once in Nova Scotia, on a calcareous ledge right here in the Corney Brook gorge. I was ready to scale a calcareous ledge if I must.

Mary and I were barely started on the trail when we made our first interesting discovery. This was a waist-high shrub called Shepherdia or rabbit berry. It is the only representative in Nova Scotia of the family that includes the autumn olive and the Russian olive, which makes it interesting. Its best feature is its thick leaves, dark green on top with fine coppery or silvery hairs underneath.

At first, the trail climbed rapidly. The river was far below, deep, dark and full and swirling around big boulders in the stream. If there were dripping cliffs or calcareous ledges on the other side of that, they were safe. It looked like good fishing, though.

The uphill side of the trail was very steep, with large old trees, and in the steepest parts, big rockslides. This slope was just as im-possible to explore as the one across the river, except for a billy goat.

It didn't matter. We were finding lots of nice things down by our feet on either side of the trail. At the bottom of the slope were big patches of partridge berry bright with last year's berries. Apparently partridges were not plentiful last winter.

Alternating with partridge berry were similar-sized patches of twinflower. These were in full bloom, with their tiny pink pairs of bells. The leaves of bluebead lily were big and luxuriant along the path,

Figure 132: Partridge berry with berries

the plants finished blooming, the berries not yet blue.

The trail began to descend and came closer to the brook. The valley flattened out and the trees were old and very tall. Every kind of fern grew among the trees. There were patches of nodding trillium that had finished blooming and were ripening their berry.

In places it was possible to get to the edge of the water. We saw lush patches of oak fern and beech fern beside the brook. Among the ferns was baneberry in bloom—hard to tell if it is the red one or the white one until berries ripen.

Near the water we found the purple avens, something we hadn't seen before, and a plant with very large maple-shaped leaves that we couldn't identify.

Back up along the path again, we came across a group of plants we rarely see around home. These are the Pyrolas.

Pyrolas are low and spreading, with evergreen leaves, and spikes of pink or white flowers; quite showy in a small way. As a group they are called wintergreens, though the true wintergreens

Figure 133: Pyrola

that flavour mouthwash and chewing gum are in the blueberry family.

There were patches of two different Pyrolas side by side along the trail. The first group was taller, maybe five or six inches including its flower spikes. The flowers were opening and seemed pink or pinkish white. The leaves were shiny.

The second group was shorter, the leaves not so shiny, and the flowers, which appeared to be white, not quite open. We tentatively identified each, using our *Newcomb's Wildflower Guide* (I'll have more to say about wildflower guides later on in the book). The first, we decided, was the pink Pyrola, the second, small or lesser Pyrola.

I was ever-hopeful of finding a new orchid, my focus for this year, and I almost did, I think. We found one small, orchid-looking plant that was about to come into bloom, but we couldn't wait around until it did. Without flowers, we couldn't be sure what it was.

By now, we were at 3 km., according to a marker by the trail. The falls were said to be at about 3.5, so we pushed on.

Sure enough, there was a falls at the end of the trail. It is not a large falls, not what you would call a cataract, but it is big enough and pleasing to the eye.

I have learned to bring along my binoculars to scan inaccessible calcareous ledges and dripping cliffs, so I trained them on the falls and surroundings in search of the fragrant fern or maybe an orchid. I have noticed this before, so I shouldn't have been surprised, but through the binoculars even ordinary plants look huge and unfamiliar. I had to recalibrate my brain before I was able to ascertain that none of those magnified plants were orchids or fragrant fern.

Still, the site was lovely, and we rested until we decided to return back down the trail.

We never did find the ledges or dripping cliffs we had read about. I think they were on the other side of the river. How would we climb them anyway?

We had a good day and were content to know that the Lapland Rosebay was up there somewhere.

There are almost unimaginable numbers of waterfalls in Nova Scotia. The fact that the province is narrow, and high through the middle, means that streams originating in the highlands have to get down to the sea in a hurry, and falls are inevitable.

Wally Ellison, in his book *Waterfalls—Cape Breton's Hidden Treasures*, writes that he has been to over 100 waterfalls in Cape Breton, and is aware of at least that many more. His book is as much a gallery of beautiful photographs, including aerial ones, as it is a guidebook.

Some of his photos are of the mossy ruins of old saw, grist, and carding mills of the past century, when almost any fast-flowing stream in Cape Breton turned a mill wheel. Ellison states that in 1851 there were 75 grist (flour) mills, 30 sawmills, and 7 woollen/carding mills working in Cape Breton. Products of these mills provided well for early settlers until goods became easier and cheaper to import.

Canada's mill wheels today, I think, are mostly turning in China.

Figure 134: Waterfall at an old mill

In the most up-to-date book on waterfalls of Nova Scotia, author Benoit Lalonde writes that he has visited over 600 waterfalls across the province and has that many more to investigate. Both he and Wally Ellison point out that there are surely a number of waterfalls they don't yet know about.

They also agree that trails to waterfalls, and the falls themselves, mostly located in gorges and ravines that have never been logged, are rich repositories of ancient trees, ferns and wildflowers.

Given the relatively small size of our province, and the incredible number of waterfalls, I wonder if we might be in contention for a world record of waterfalls per square mile. If I ever find all the native orchids and all the native ferns of Nova Scotia, I may switch to waterfalls.

Figure 135: Black Brook Falls

10: National and Provincial parks and trails

Figure 136: Camping with the Beetle

Each summer, Mary and I shine and tune up our '72 Volkswagen Beetle, and head out onto the road for a week-long trip to some interesting part of the province. We stuff every cranny of the old car with camping gear—tent, sleeping bags and pads, collapsible chairs, cooler, water jug, camp stove, spare clothes, an assortment of footwear. You get the picture.

The choice of the old VW for the trip is a good one. Everyone in Canada owned one at one time or another, or had a friend or a girl-friend who did, or their parents or brother or sister drove one. There is palpable public nostalgia for the Beetle, which, during its heyday in the 60s and 70s, meant happy times.

When we pass through town, people smile and wave, or toot their horn or blink their lights. They don't even have to see us; they hear us coming—the unmistakable rattle of the old Volkswagen air-cooled engine. Even little kids, who must have seen a Beetle in the movies, jump up and down with glee and give us the thumbs-up.

When we pull in for gas or groceries, people come over to look at the car and chat. They tell us about the one they drove or how when they were tiny kids they used to ride in the empty hatch behind the back seat. Yes, it gets people talking. When we drive into a camping spot to set up our tent, younger people are astonished to see us pull our camping gear out from under the front hood, where the motor is supposed to be.

I can't say enough good about Nova Scotia's provincial parks and the people who work in them. Our routine is to pick a park that looks like we could be there by afternoon to set up camp, and phone ahead for a reservation. The park person on the phone is always extremely helpful, asking if we want a site that is treed or open or if we want a site by the water. When we arrive, if we don't like the site we have reserved, they will help us find another one. Often they tell us which site they think is the best one in the park, and help us get it.

The first driving tour we took in this way was a four night, five day drive completely around the Nova Scotia shore, during the height of the COVID-19 pandemic in 2020. In the parks, every second campsite was closed to allow for social distancing.

Oddly enough, during the pandemic, people took to camping in droves, a completely unexpected but wonderful phenomenon. It was not easy to get a campsite. Sometimes we hadn't decided until late in the afternoon where we would stay for the night and were late phoning for reservations. Again, though, the park staff were great and always found us a nice spot.

This was the first time we had sampled so many provincial parks, and we discovered that virtually all of them boasted interesting hiking trails. This trip, however, was mostly a driving one.

Driving an old, four-speed, stick-shift Beetle was plenty of fun, but didn't leave much time for hiking or hunting for wildflowers. We pulled in to campsites early in the afternoon and took off again after breakfast in the morning. Reminiscing at home after the trip, we agreed that the next time around we would visit fewer parks, stay longer, and try those trails.

I would like to tell you just one more story about that trip before I get back to talking about wildflowers, because it is a story of how good people, and especially Volkswagen fancying people, can be.

We had driven from Digby all along the French Shore to Yarmouth, enjoying it all the way, and camped that night at Ellenwood Provincial Park. The next morning we kept on past the Pubnicos and the estuary of the Tusket River—the most fabled river system for rare wildflowers in Nova Scotia. We had little time to explore, but promised ourselves to come back someday.

We continued along the South Shore on the little roads, and were enjoying ourselves and making good time when we blew a tire. This was not an immediate problem because we had a good spare and I put it on.

The tires we had started out with were the tires that were on the car when I bought it, and may have been on there since 1972. When I examined the blown tire, the tread looked good but the sidewalls were badly cracked and had burst. Maybe, I thought, I should have a look at the other three tires on the car. Well, they too had good tread but cracked sidewalls, just like the one I took off

That's when I panicked. We were still several hundred kilometres from home, had used up our spare tire, and had three other tires that were no better than the one that blew. Chances were very slim, I thought, that we would get home without another blow out, and with no spare.

We left the winding shore road, got on the fast highway and pulled off at the next town, which I think was Liverpool, to buy another tire for a spare. Glory be, there was a tire shop just after the exit. I parked and went in.

Before I could say a word, the old tire man at the desk, without looking up, said, "They don't make tires that size anymore."

Taken aback, I asked him what did he mean and couldn't there be something in his stacks of tires that might work. He shook his head and said he would phone other tire shops, which he did with no luck. One of the shops, he said, had taken their old VW tires to the dump just a week ago.

He did give us the number of the Old Volks Home, a garage in Mahone Bay that specialized in antique Volkswagens. He hadn't been able to get them on the phone.

It seemed to me that the Old Volks Home would be sure to have a tire, and we sat in the car trying unsuccessfully to get through on our cell phone. Meanwhile, the old fellow in the tire shop came out to tell us that one of his men said there was a house in Liverpool where he had seen an old Beetle up on blocks, and told us where to find it.

The house wasn't hard to find, and, sure enough, there was an old VW there, but it was back together and wasn't up on blocks. We knocked on the door, which was opened by a nice young woman who listened sympathetically to our story, but was afraid they had no extra tires to sell us.

We sat in the car, and finally were able to get through to the Old Volks Home. They told us that they could order a tire but it would take several days to arrive.

We were on the point of driving off to take our chances on the road, when the young woman came out of the house again. She had been on the phone, she said, and had found us a tire. Some friends, who also kept antique Volkswagens, had a tire on a rim they would sell us. The guy had said it was old but still held air.

Well, this was good news. We asked for directions, and the woman said don't worry. She and her husband, who had just come home, would lead us there.

So, they drove and we followed. At the place they took us to, the man and his wife had a beautifully restored black Beetle and a blue VW window van. They also had the tire and rim, which they only wanted twenty dollars for.

We paid them and I tried to pay the couple who had guided us there, but they wouldn't take it. We thanked them all profusely and hit the expressway for the provincial park where we had reserved a campsite for the night.

We will always remember the kindness of those people. We drove home the next day breathing easier, and, incidentally, never blew another tire. I immediately ordered new tires, made in Korea, and they are on the car now.

Figure 137: Kejimkujik Park

The following summer, 2021, we made for Kejimkujik Park, the national park, after camping one night at Smiley's provincial park, and visiting friends in Petite Riviere. The national parks, of which there are two in Nova Scotia—Kejimkujik and Cape Breton Highlands—are spectacular, with good campsites and over-the-top comfort stations boasting dishwashing facilities and lockable, private, sound-proofed cubicles with electrical outlets, a toilet, a sink, a shower and a mirror on the wall in each one. Even the firewood they sell for five dollars a bag is dry.

Firewood is dry in the provincial parks, too, these days. Not many years ago, all they sold was damp softwood that smoked but wouldn't burn. We always packed our own firewood from home, but that isn't permitted anymore due to worries about invasive forest pests, so campers must buy it at the park.

It is a testament to the integrity of workers in the parks that, since you are forced to buy firewood, they will make sure that it is good and dry.

The Kejimkujik National Park is best known for camping, canoeing, and kayaking on the big lake; but for the landlocked, there are great hiking and nature trails. The trails are well marked, well laid out, and used by hundreds of visitors each summer.

The park website lists 12 frontcountry trails, from 1 to 6.3 km. in length, through various types of forest and topography, some following the beautiful and historic Mersey River. In addition, there are two backcountry trails for experienced hikers, the Channel Lake trail of 24 km. and the Liberty Lake trail, 54 km. one way. In his book *Hiking Trails of Mainland Nova Scotia*, Michael Haynes describes the two long backcountry trails, and six of the frontcountry trails in detail.

Though Kejimkujik Park lies largely in a region of wetland and lakes, the common

Figure 138: The Mersey River

denominator is actually majestic forest, some mixed and deciduous, and some old growth hemlock and pine. Mary and I were there in September, as summer was winding down and not much was going on under the trees, but there is every reason to believe that the wildflower show would be good in the spring. The National Park trailbuilders have worked their magic in Kejimkujik, and have built weeks of good hiking for those who can find the time.

Kejimkujik Park is divided into two pieces, the main park around Kejimkujik Lake and a small piece on a peninsula above the seashore near Port Joli, called the Seaside Adjunct. The trail through the Adjunct is said by Michael Haynes to be one of the busiest and most popular walking destinations in Nova Scotia in summer.

The wide and nicely gravelled trail passes through dense, shoulder high, seaside shrub growth including bayberry, alder, birch and mountain ash, with glimpses of seashore bogs and barrens along the way. At its far end it descends to a spectacular white sand beach with lounging seals.

Mary and I and our daughters took a quick walk on the trail to the point where we could see the beach. It was at the end of the day, because it had rained until early afternoon. This was August 28, and there weren't many people there that day.

Figure 139: Nodding ladies'-tresses orchid

I found my last orchid of the season—or I should say Mary found it for me—along the trail in an open wet spot: the nodding Ladies'-tresses.

The other national park in Nova Scotia, Cape Breton Highlands, is another masterpiece, and they are constantly making it better. Like Kejimkujik, it has earned a solid reputation worldwide as a destination for outdoor adventure, and has even more hiking trails, some long, some short, some easy, some difficult.

Much of the attraction of the park is its spectacular cliffs and mountains that drop straight into the sea. Trails rated difficult are difficult—steep, that is, but not rough.

Trails in the park, though heavily used, are constantly inspected and maintained, and provided with seated overlooks and benches to make any hike enjoyable. Hikers are instructed to stay on the trail, and, and they do, so wildflowers and ferns alongside are untrampled for Mary and me.

We visited the Cape Breton Highlands park a couple of times last summer, and hiked a few of the trails. I have already described most of these: the Skyline Trail, the trails to Corney Brook falls and Benjie's Lake, and the boardwalk around the orchid bog.

Years ago we descended the long (6 km.), steep trail to Fishing Cove, inappropriately equipped with heavy tent and packs and

Figure 140: Cape Breton Highlands National Park

sleeping bags on our backs. The trail down was lovely, with lots of interesting plants as well as springs and waterfalls splashing down alongside. The weather was very warm.

At the end of the trail, where the Fishing Cove brook runs into the sea, we set up our old tent in a beautiful meadow. We watched the sun set as we ate our supper on the beach, followed by a campfire. Things don't get any better.

They went downhill from there.

We found out why hikers didn't camp with canvas tents like ours anymore. First, they are heavy. Second, the screen across the doorway kept out mosquitoes, but no-see-ums went right through it. It was a hot night and it was impossible to close up the tent without roasting. We lay unclothed on top of our sleeping bags and were attacked.

We had never really encountered a serious assault by no-see-ums before and it took a while to figure out what was biting us. It was dark and we couldn't see um. The bites itched and burned something fierce and we were forced, finally, to spend a miserable night sweating in the heat with our sleeping bags pulled up over our heads.

In the morning the no-see-ums quit biting. As the sun shone into our tent we saw the peak black with millions of the microscopic little devils.

Once we were up and dressed we ate breakfast, hiked and swam and forgot the bad night. In the afternoon, we packed up our heavy load and headed back on the trail—6 km. again, all uphill.

That canvas tent almost found itself flung into the ravine a few times, but I was fond of it and carried it all the way to the top. We still have it, but only take it car camping.

Other trails we have tried in the Highlands include the Acadian Trail, up a steep series of switchbacks to the very top of the highland plateau where the firs and mountain ash are all gnawed by moose. We went up that one when our daughters were small. I carried the younger one most of the way. The trail was so steep that the older girl went up on all fours.

One other trail we have tried, Salmon Pools, goes up the Cheticamp River, where we saw the little bluebells of Scotland blooming in every crack of the rocks.

We have more trails to try In Cape Breton Highlands Park and more wildflowers to discover, so we will be back.

There are two national parks in Nova Scotia. There are over one hundred provincial ones, twenty of which are set up for campers. The rest are what are called picnic parks, which close at night. Almost all provincial parks are on water, be it a lake, a stream, or the seashore, and nearly every one includes hiking trails, which means wildflowers.

Provincial parks are less glamorous than national ones, but well laid out, comfortable, and suited to the average camper and their families.

The elegant, architect-designed visitor centre of the national park, with gift shop and bookstore and uniformed staff behind the sweeping counter, is more likely a log building, with a local girl printing out passes on a plywood counter and a rack of used books to pick from for free at the provincial one. Two or three park workers, men and women, run between the office, where they help new

Figure 141: The provincial parks of NS welcome you

arrivals and sell ice, to the woodpile, where they measure out and tie up five dollar bundles of firewood.

For most of us, camping is more than just setting up a tent and sitting by the fire. The camp is our little home, but we come to swim or fish or hike as well, and park planners have wisely provided for all these activities. In the provincial parks, trails are established, of every length and degree of difficulty, to suit anyone. Some trails take you to beaches, some to scenic look-offs, and others make loops through interesting terrain, and inevitably, past wildflowers.

In his hiking-trails book, Michael Haynes singles out Taylor

Head, Blomidon, Thomas Raddall, and Cape Chignecto parks for exceptional hiking, though I know for a fact there is good hiking in many others.

Also, wildflower exploration doesn't have to be a hike. Sometimes a saunter or walk better describes it, and a short trail will do. We admit, though, that there is a certain satisfaction in completing an ambitious hike.

We haven't tried Taylor Head or Thomas Raddall, but we can vouch for Blomidon and we tried a little bit of Cape Chignecto.

There are good campsites at Blomidon, but I would advise you to pick one among the trees. We started with one in the open field with a spectacular view, but the wind came up and nearly blew us off the cliff. We moved to a nice wooded spot.

The next day, Mary and I and our daughter Margaret hiked the trail from the up-

Figure 142: On the trail at Blomidon Provincial Park

per campground to the end of the cape, the home of Glooscap, where one can gaze across the Minas Basin to the red cliffs of the five islands. Along the way were several wooden viewing platforms situated at incredible look-offs. Seagulls, ravens and hawks flew below us along the cliffs.

Cape Blomidon has long been known as a special place for native flowers and ferns. In 1882, George Lawson Ph.D., professor of chemistry and mineralogy at Dalhousie College and University, Halifax, and an enthusiastic botanist, wrote:

One of the most attractive features in the scenery of Nova Scotia is the bold and strikingly picturesque promontory of Blomidon, rising to 400 feet in height, which forms the north-eastern termination of the North Mountain, and now looks down upon the fertile stretches of waving meadow, blossoming orchards, and scattered towns and villages, as it did in the olden time on the less ambitious hamlets and carefully cultivated fields of the French farmers. The phys-

Figure 143: Red trillium, leaves and berries

ical and geological features of Blomidon—its red sandstone strata, mostly covered by a debris-slope, and its continuous summit cliff or wall of dark trap—have often been depicted by pen and pencil, and its zeolites and other treasures of mineral species are shown in most of the public museums of America and Europe. It is not so well known that Blomidon is a rich pasture for the botanist.

Prof. Lawson goes on to elaborate on the rare flowering plants and ferns he and his companions found on the promontory of Blomidon.

140 years later, we were ourselves impressed by the number and variety of ferns, and especially by the immense size of the trees, both evergreen and deciduous. We were at Blomidon Park in late July, when not much was blooming on the forest floor, but I did note numbers of an unfamiliar trillium which I identified by elimination as red trillium, which I had never seen before. It wasn't the nodding one, and it wasn't the painted one, and there are only three species of trillium in the province, if you don't count an isolated discovery of white trillium, so it had to be the red one.

I have since confirmed the correctness of my guess because two hiking books describing Blomidon both mention the "carpets" of red trillium in June. Needless to say, we are anxious to return when June comes around again.

Our Cape Chignecto adventure was part of a four day, three

Figure 144: Tide out, Five Islands Provincial Park

night camping trip to Five Islands provincial park that began September14. This park overlooks the Minas Basin, directly across from Blomidon.

Five Islands is not mentioned in Michael Haynes' hiking book, but should be. There are some good trails there. This park is famous for walks along the beach below the red cliffs, when the Fundy tide is low and the Minas Basin is empty to the bottom. At low tide it is possible to walk across the muddy ocean floor to an offshore island if one comes back before the tide turns. If not, it's a twelve-hour wait on the island until the basin is empty again.

The Fundy tides are simply hard to believe. At the park there are plywood clocks set to show the time of the next tide change and warnings to not get caught on the beach or halfway out to the island. The tide floods in fast, and what was bare mud a couple of hours ago becomes brim-full ocean. I had to wonder if the fish and sea creatures swim out of the Minas Basin and then back in again, twice a day.

Mary and I did our walk on the beach and out to tall rocks. Both the beach and the rocks disappeared without a trace when the tide came in. After that, it was time to try drier trails, which took us along the top of the bluff.

We selected the Red Head trail, which went for a little over two kilometres and took us along the top of the bluff and eventually to a wonderful look off at Red Head. Along the way, we passed through some dense coastal coniferous forest and then into patchy young mixed forest where there was evidence of past settlement—abandoned fields growing in with brush, and ancient apple trees. We found old-fashioned plants such as tansy and valerian, which probably grew around the vanished farm house.

Farther along, the trail entered old and beckoning deciduous woods where we left the trail for a bit to explore, but at this time of year found mostly ferns. There was, though, in this area and places along the trail, an orchid. It was the broad-leaved helleborine, and there were lots of them, some still in bloom. This is the orchid that Carl Munden calls the weed orchid, and I could see why. There were hundreds.

This wasn't the first time I had seen this orchid, and I was getting tired of it. The flowers are small and greenish, or slightly read in a spike, and not particularly beautiful. If it was nicer-looking, we might not call it a weed.

The trail continued on over a series of small bridges and ended at Red Head, where we rested a while and enjoyed the view. On the way back, rather than retracing our route exactly, we branched off

Figure 145: Red Head

onto the Economy Mountain trail, which cut through nice mixed forest for another couple of kilometres and then brought us back to camp.

The forest floor was rich and mossy with ferns and all the familiar plants. I did get excited briefly by a rugged-looking orchid that someone had cut off. I thought that maybe we had discovered a new species, but slowly realized it was just an exceptionally large specimen of the weed orchid from which someone had taken the flower spikes.

The last part of this hike was a gruelling uphill trudge on the paved park entrance road back to our camp. We flopped into our folding chairs to recover and talk about the things we had seen that day.

The highlight of our afternoon hike was the discovery of a truly ancient yellow birch, the likes of which we don't often see on the East Coast. The wide-spreading branches were hung with lichens and moss.

Figure 146: A forest on a tree limb

One enormous old branch jutted out at a right angle toward the trail. On top of it was an actual little forest. Growing from thick sphagnum moss were ferns and twining plants from the forest floor, as well as tiny trees—spruces and birches—which had somehow seeded into the moss.

How this little forest came to grow on a tree limb we could only guess, but it looked as if it belonged in the West Coast rain forest. We suspected fog.

Sure enough, the next morning we woke up to thick fog. We had planned this day to drive to Parrsboro, then keep on to the Cape Chignecto Provincial Park.

Cape Chignecto is unique in the provincial park system. There are no drive-in camp sites. All sites require packing in, and the trail is not for the faint-hearted.

The hiking trail is the longest of any in the park system, 51 km., and passes over barrens and through steep canyons and ravines. The hiking guide recommends that the loop should be done in two overnights and three days of hiking. We weren't going for three days, but we thought we might walk around a little to get the feel of it.

The fog wasn't lifting, but we set off in the old VW anyway. It was rather eerie because we could only see the road. What could have been primeval forest and elves along the road was obscured by the fog.

The road was challenging to drive in an underpowered Beetle. It was steep and winding and required lots of gear shifting. The worst was when speed signs slowed us down to 30 km./hr. on a piece of road snaking downhill to the bottom of a valley. From the bottom, heading steeply uphill again, it was impossible to get past second gear in the little car, and we couldn't pick up speed again until the crest and the next downhill, where it happened all over again.

The road was so steep and winding that there was no passing. We tended to accumulate a string of faster cars behind us and had to pull over from time to time to let them by.

We passed through Parrsboro without being able to see it, and kept on to Advocate and Cape Chignecto park. By then the fog had lifted some and we ate lunch, shouldered our packs and thought we would try a little bit of that 51 km. trail.

Soon the trail was heading steeply uphill through attractive Acadian forest. We toiled uphill for a piece, then stopped to catch our breath. Then uphill again. We hiked for about an hour and a half and were still going steep uphill and it was still uphill as far as we could see ahead.

We are not too proud to give up at a time like that, so we did, and returned to the car.

The ride back to Five Islands was enjoyable because the fog was gone and we could see. The little villages were quaint and the scenery awesome. Steep, thickly-treed mountains rise on one side of the road, the equal of any in Cape Breton. On the other side, streams of water cascade down steep ravines and plunge into the

Bay of Fundy far below. I'm surprised that this part of Nova Scotia isn't promoted more for tourism, but then, I'm glad it isn't.

The only other provincial park we visited last summer was the one they call The Islands, across the bay from Shelburne. This park was small, with no outstanding hiking trails, but its campsites were very nice, scattered among fern-covered outcroppings of rock.

From our camp, we could see, not far across the water, the town and the fisheries and boats of Shelburne. Throughout the park were leaves of lady's slipper orchid. We were there at the end of August, so the orchids had long since bloomed, but the park must be lovely in June.

Coming from Antigonish, the town of Shelburne and its environs feels like another province to us. The weather is milder and the soil is different from where we live. The vegetation is different and I expect there is a great deal of fog.

If you are nearing Shelburne on the highway at the right time of year, you can spot the bright yellow flowers of Scotch broom, a spiky-looking shrub that grows three or four feet high and is seldom found elsewhere in the province. Other rarities are the inkberry holly and the skunk cabbage, with its big leaves, and acres of bearberry. I have never been close enough to a skunk cabbage to find out if it gets its name from its smell.

The only trail we found at The Islands park was a rather tame one that starts at the park, and after a short distance crosses the Roseway River alongside the highway. Shelburne, though, is close to major lakes and rivers in the interior where there must be trails, and to the Kejimkujik Seaside Adjunct trails.

July 7-8 2023. One more good provincial park to add to the list. After keeping an appointment with Sean Blaney at the Atlantic Canada Conservation Data Centre in Sackville, N.B., we camped two nights at the Amherst Shores Provincial Park in Nova Scotia.

This isn't far from Amherst and is a comfortable campground close to a wide beach of red sand. On the weekend the campground was full. The days were hot and I think virtually everyone had come to cool off at the beach.

Mary and I discovered that the park boasted some remarkably interesting trails that no one seemed to be using. The trails are intersecting loops passing through mature Acadian forest, which was badly damaged in places by Hurricane Fiona, but has been cleaned

Figure 147: Partridge berries at Amherst Shores

up by Parks staff.

These trails are a combined total of four kilometres and are not difficult. In large part they follow the fern- and flower-lined shore of the tea-coloured Annabelle's Brook, and appear to be very lightly used.

The forested parts of the trail shelter a very interesting understory of flowering plants and ferns. There were areas where there were many lady's slippers and painted trilliums that were finished blooming. We also found Pyrolas, partridge berry in bloom, and patches of the curious cucumber root with whorled leaves and drooping flowers.

The brook was densely bordered with ferns and various tall flowering plants, including meadow rue blooming more beautifully than I have seen it other places. Everywhere the trees were majestic and shady and I think it was more comfortable in the woods than it would have been at the beach.

The trees were also full of an abundance of birds. We identified a couple of woodpeckers and the white-throated sparrow. We saw robins and heard the hermit thrush. In slow parts of the brook were cackling kingfishers plunging into the water for minnows. We glimpsed many more birds that we couldn't identify. I think this park is also known for massive flocks of shorebirds that stop there on their fall migration.

When Mary and I go looking for wildflowers we carry binoculars to watch birds. At Amherst Shores Park we stumbled along trying to look up in the trees and down at the ground at the same time.

Figure 148: Indian cucumber root

There are numerous other provincial camping parks that we would love to visit. All have trails and wildflowers. They are scattered across the province from one end to the other and encompass all sorts of ecology and terrain.

Besides the camping parks, I tried to count the day use parks, many of which have hiking trails, too. There were over a hundred. One is the Pomquet Beach Park that I have already raved about in the chapter on mayflowers. Another is the Port Shoreham Beach Park from which I nabbed a photo of the sign.

The province may be stretched thin for health care but they are doing a nice job on the parks, which, for the happy families swim-

ming at the shore, or strolling the park in the evening, or riding bi-
cycles in the open air with no traffic to run over them, is itself the
finest of health care.

Chief Charlie Labrador

1932-2002
6th Generation Mi'kmaq of Kejimkujik
Mi'kmaq Elder and Spiritual Healer

If we could see beneath the forest floor, we would see that all trees,
the pine, birch, maple and so on, are holding hands,
regardless of species.

We as people, regardless of race, must come together and hold hands
and help each other.

We must listen to our Mother Earth.
She is sending us a message.

Charlie's spirit will watch over Kejimkujik for all time.

Figure 149: Chief Charlie's wisdom

Figure 150: The great outdoors

11: Getting outdoors

Figure 151: Ready for a ramble

Most of the time, for most of us plant hunters, day trips are the rule; but sometimes we are lucky enough to spend a few days away from home in the wilderness, or the country anyway, hiking or camping and botanizing. Extended trips into the countryside require planning.

As an amateur botanist, there are some things that you must

take with you. You shouldn't need 5000 sheets of drying paper and 16 heavy freight boxes, like M.J. Fernald brought with him in 1920, but you could start with a comfortable day pack.

The need for a light pack is obvious if you are out for the day, because you will want to take water and some food. Also some dry socks and some extra clothing, depending on the time of year. Beyond that, here are the essentials:

1. A map and compass

Have a look at a map before you start out, so you will know the approximate compass heading to follow going in and which to come back out on, or take a reading on your way in and add or subtract 180 degrees to get out. This, of course, is less essential if you are following a trail or a stream or a lakeshore, or if you are in a small open area where the visibility is good, unless fog or fading daylight catch you.

When away from the trail, wandering around following your nose and going from one patch of vegetation to another, you can get hopelessly lost. I know, I have done it more than once, and it is not a good feeling.

It's true too, that when you are wandering around lost, you go in circles. I was lost once hunting rabbits in the snow. I walked and walked, sure that I was heading out, only to come upon my tracks again, heading in the same direction. Trying not to panic, I struck off in a new direction and was lucky enough to find a woods road before I did a circle again. That is when I began to carry a compass.

Even without a map, if you have been paying attention, you should be able to make a good guess as to which direction to take to find your way out. If you are off a bit, following the compass enables you to go in a straight line instead of circles, and you will come out somewhere.

The most important and ironclad rule is to trust your compass. It is very common to think that you know the way out and so the compass must be broken. That's why you are lost.

René J. Belland tells the story of an expedition with the well-known Nova Scotia botanist W. B. Schofield, and getting lost in Cape Breton:

> Once down inside the gorge we collected numerous rarities, but soon noticed that our botanist guide had disappeared. When he did not reappear after three hours we decided that we had been abandoned (and indeed we had been!). It was late afternoon when we made our way back to the rim of the gorge, but there were no bogs to follow back to the road, only dense coniferous forest. At once, I retrieved my compass and map and calculated a heading through the woods to return to the car. I soon discovered Wilf's mistrust of mechanical devices. Five minutes after starting out, he states that the compass is wrong, and that we should head left. I replied that I had done this before and that trusting the compass would get us home. Another five minutes passed and Wilf again says, this time more fervently, that we're heading the wrong way. I retorted with, 'I'm following the compass. If you want to go that way, then go ahead'. No reply. But I could hear Wilf following close behind as we crashed through the bush. Another twenty minutes passed and we finally reached a large bog. On the other side we could see the car, smack on our compass heading.

In 2023, I suppose, the GPS makes the compass redundant, which is fine, unless you drop it in the water or the battery goes dead. It's good to have both.

2. The next thing to go in the pack should be a **good knife and matches**. Maybe you can carry the knife in your pocket or on your belt, but keep your matches in your pack and dry. You might carry a little newspaper for kindling, though a bit of birch bark will get the fire going even in wet weather. I think you can see the usefulness of the knife and matches, even if it is only to light a fire at lunch time to warm up by. And if you do get lost, well....

When I am out in the boondocks, I always carry my old Felco pruners in their pouch on my belt. They are good for cutting any-thing—even wire or tin—and can also be used as a knife. I fre-quently wear them around the house and yard, and it's amazing how often I am pulling them out to cut something. I'm not trying to sell Felco pruners, but they are good to have, especially if you are a gardener or forager. Felco are the toughest.

I am re-reading this bit about matches and knives. It is June 2023, the worst fire season the province has ever known, and even the thought of a small cooking fire makes me nervous. Maybe it is better to save the matches only for emergency, and build your fire in the fireplace back at camp.

3. Bug and tick repellent
Biting insects or ticks can really make a mess of a good day out-doors. There are many brands of repellent for black flies and mos-quitoes available these days, and you can try them and decide which you like best. I think they work in the case of no-see-ums, too, the worst biting fly of all. Unless you have very fine screen in your tent, no-see-ums can get through.

Fortunately, no-see-ums don't seem to be as widespread as black flies and mosquitoes. Horse flies and deer flies drive cows and horses crazy and can be annoying, but since they are mostly interested in the top of your head, wearing a hat pretty well stops them.

I have to confess a grudging respect for the black fly. It is the gentleman of biting flies. It plays by the rules. It may attack in over-whelming numbers during daylight but, unlike the cursed mos-quito, it doesn't bite after dark or indoors. Also, in most places, it is finished and has disappeared by mid-July.

I should mention that the mesh bug shirts that are sold every-where now are very good. Sometimes, in some light, they do ob-struct your vision, but if the bugs are bad enough, we can put up with that.

Try zipping up your bug shirt, then dropping on a wide

brimmed hat over that. This keeps the sun off your face which makes it much easier to see. I wear mine when working in the garden, seeding small seeds, where I would otherwise be a sitting duck for flies. The shirt shuts me off from the bugs so effectively that I have to unzip it once in a while to see if they are still there.

Ticks are in a league with cockroaches and rats and tse-tse flies and all those things Noah should have left on the ark. The fact that they are creepy and dig under your skin to suck the blood is not enough. They also carry Lyme disease, which I and some of my friends have contracted at one time or another.

It is hard to make sense of where you are likely to run into ticks. I think the worst is where it is brushy or where there is high grass. Sometimes, though, you spend the day in a place like that without any trouble, then pick up a tick going down to the garden to get a tomato.

There are sprays these days that claim to repel ticks, and if you anticipate tick trouble I would certainly try those. Some of my friends wear overalls soaked with permethrin, which kills ticks on contact, but treated overalls are uncomfortable in hot weather. The best defence seems to be light-coloured clothing with a tight neck and cuffs, and socks pulled up over your pants legs.

I don't think that you should worry so much about ticks that it spoils your peace of mind. Just check yourself over good when you get home. Usually you will find nothing.

Buy some sort of mechanical tick-removal tweezers and if you do find a tick you can pick him off. They say the tick has to be on for 48 hours before it can pass on Lyme disease.

A curious discovery that we have made is that ticks seem to get under your clothing while you are outside, then stay put. In the evening, while you are sitting quietly, they begin to move and you can feel them tickle. Mary and I are now attuned to the feeling and never ignore a tickle. Wait until you feel it twice, though. Don't be whipping off your clothes for false alarms.

Lately we seem to catch ticks before they ever bite. I am out-doors a lot, and my tickle take for the last couple of weeks has been one on my forehead, two on my ears, one under the sleeve of my shirt, and one crawling up my leg in bed.

Before I get too smug, I'd better knock on wood.

4. A light-weight first aid kit

It is usually the band-aids that you need, so make sure there are enough of them in various sizes. Small commercial kits include other things you might not think of, such as tweezers and bottles of aspirin, and are probably worth buying.

I think these are all the essential things that you really have to have in your knapsack, but here are some others that are optional but can be nice to have:

5. A field guide to wildflowers

There are quite a few good field guides now for wildflowers of the Northeast. You want it to be compact, so don't buy the complete wildflowers of Canada or North America.

I like the field guides that use paintings and sketches better than those that use photographs. With sketches and paintings, the artist is able to illustrate specifically the features that most clearly identify that species. Sometimes a photograph is from a slightly bad angle or poor light and you can't quite see what you want to see.

I have an old *Peterson Guide to Wildflowers of the Northeast* that I like, and we recently bought *Newcomb's Wildflower Guide* to go with it. I was in the library the other day and saw others that looked good.

We sometimes go exploring with no book at all, but we carry a plastic bag to put specimens in. If we encounter a plant that we don't recognize, we snip off a piece of stem with flowers or fruit, if possible, and a few leaves. This goes in the plastic bag to keep it from wilting, and we use the guides and the Flora to identify it at home.

We don't take roots with the specimen, and we won't take a

piece of stem if there is only one plant or if we suspect it is rare or endangered. We don't want to kill the last one in Nova Scotia.

6. Hand lens and binoculars

In plant taxonomy class at university, students are all equipped with a hand lens for examining flowers. A hand lens is a tiny magnifying glass that you hold close to your eye to count stamens or pistils. Mine is 10X power. I take it into the field but often don't use it until I get home and take my specimens out of the bag.

For me, at my advanced age, binoculars are indispensable on any trip into the outdoors. I use them to scan cliff faces and steep slopes or likely-looking spots in a bog to determine what is growing there. If something looks interesting, I might climb or wallow closer to have a better look.

The only problem is that everything looks interesting through binoculars because of the magnification, but you soon get used to that. And on top of all this, binoculars are good for watching birds and wildlife that you are sure to come across while plant hunting.

7. A small notebook and a pencil are very good to have in your

bag. It is surprising how often you want to write something down, or to make a little sketch. My daughters gave me a clever little notebook that can even be used in the rain.

8. Bear bells and bear spray

Bears are something we never used to worry about. There seems to be a large population of them around Antigonish County these days, however.

My theory is that the bear population began to explode when, in recent years, farmers began to grow enormous fields of corn for cattle feed. The bears are fattening on the corn and raising large families.

I actually walked between the rows far into a corn field one year, and found spots as big as a room in a house where the ears of corn were stomped down and half eaten. The ground was too hard to find tracks, but it seemed that it could only be bears doing something like that.

One day, the farmers cut down and harvested all the corn, and, sure enough, on my way into work the next morning there was a big bear at the edge of the field looking grumpy and wondering where the corn went.

Since there are more bears, we put a cylinder of bear spray into a pocket of the pack where it will be easy to grab, then fasten bear bells to the straps to jingle and prevent coming upon a bear by surprise.

Bears, as a rule, are as timid as deer and we hope, of course, never to have to use the spray because it is very nasty stuff. It is reassuring, though, to know that we have some defence if we need it.

9. Finally, by all means, take your **cell phone or camera**. You will certainly want to take a few photographs, or text for help if that bear spray doesn't work. Also, a photo of a rare plant that you don't want to pick and take home for identification will help you remember what it looked like.

I have recently learned of the iNaturalist app for the cell phone that will help you identify anything you take a picture of, including animals, birds, snakes, reptiles, butterflies, even lichens and insects. Almost anything in nature in other words.

I write more about iNaturalist in chapter 13.

12: Etiquette in the wild

Etiquette in the wild is simply consideration and good manners.

It goes without saying that we shouldn't litter, or cut down trees or cut branches from them. It is naughty to steal or deface trail signs. We don't paint or scratch our names on rocks or carve them into trees. Don't do caca right next to the trail, same goes for your dog, and bury your toilet paper or put it under a rock. You won't often need toilet paper in the wild, anyway, because you are never far from damp moss or soft grass.

Stay on the trail as best you can. Try not to take alternate paths that may cause erosion. If you want to get off the trail to do some looking, be careful not to knock down rocks or mud, and watch where you step. Go slowly and don't squash a rare orchid. In most cases, it is a no no to get off a boardwalk, so don't bother.

Don't build a fire unless it is in an official fire pit provided by the parks or campground owners. We have seen the horrible consequences of a fire getting away.

As far as the herbaceous plants go, you may pick flowers in fields, along the roadsides and ditches, but not in the forests and bogs. We confess to gathering fern fronds in the forest where they are plentiful, and sometimes a stem of a flowering plant for identification purposes.

Never pick orchids or lilies if you find them, and don't dig up plants for your garden indiscriminately. Find out what they are first, and if they are plentiful, and if they tolerate being transplanted. If they are on crown land, or you have permission from the landowner, you might take one or two, but don't leave holes. Fill them in. Orchids can't be transplanted, so don't try.

Private property is an important issue, and you really must be respectful. Trails through private property are especially vulnerable. If a few people crossing his land are misbehaving, the landowner has the right to close the trail to everyone.

Trails to many waterfalls in the province pass through private land. In most cases the landowner accepts this with generosity and good grace, but it doesn't take much to upset the applecart. Imagine if it was your property and people were littering and painting their names on the rocks. You would probably close the trail.

Landowners with waterfalls or dangerous steep terrain on the property have more to worry about than litter or vandalism. Allan Dillard, in his book *Waterfalls of Nova Scotia*, explains it in these terms:

> It is not the usual issues of litter, trail erosion, and environmental damage which most property owners resent, however. It's the visitors who are even more thoughtless; the ones that might push reasonable safety concerns beyond their limit and risk personal injury. Those are the hikers who bring landowners to the point of barricading their property. When their land is home to a wilderness waterfall site, landowners rest nervously, due to the great dangers associated with deep gorges, faults, fissures, slippery rock faces, and litigious visitors.

In other words, behave yourself, go cautiously, and don't expect to sue if you get hurt. Legislation now protects the landowner, giving anyone hurt on his property no more rights than an illegal trespasser.

Allan Billard also counsels us to ask the landowner's permission if the trail passes near his house, or a sign asks us to.

His final bit of advice is to leave the property not only as good as you found it, but better! Take a garbage bag with you in your knapsack. Pick up your own garbage on the way out, and any that you might find along the trail.

Let's give those good landowners no reason to barricade the trail on us.

13: Identifying your finds to protect biodiversity

Figure 152: Protected at last!

The time seems to have arrived for native plants to take their rightful place in our province—in the world, for that matter. In view of increasing population, traffic congestion, noise, extreme weather events, and threats to the environment, everyone senses the importance of saving and protecting what we have. That means all wild creatures and habitats including wildflowers and native plants.

The provincial government, which not too many years ago seemed to believe that wild areas only stood in the way of progress, has made an about-face. The Department of the Environment has become very protective of native plants and rare and endangered species, to the point that we have seen endangered plants stop golf courses. Commercial and industrial construction projects must meet rigorous environmental impact assessments, which involve lengthy studies that may cause delays or scuttle the project altogether. Concerned citizens and amateur naturalists keep a close eye on developers and developments as well, and don't let them get away with anything. Times have changed!

To further ensure the safety of native and endangered species and habitats, the N.S. Department of the Environment is designating more and more sensitive habitats as Nature Reserves under the Special Places Protection Act, or Protected Wilderness under the Wilderness Areas Protection Act. Within these areas, all destructive activities such as wood cutting, fires, camping, motorized vehicles, hunting and trapping and littering are prohibited.

Between 1981 and 1984, the first seven Nature Reserves were established in Nova Scotia. In 2023 I count 100 Nature Reserves and 75 Wilderness Areas currently protected in the province.

Sensitive natural areas are selected as Wilderness Areas or Nature Reserves by the Protected Areas branch of what is now known as the Nova Scotia Department of Environment and Climate Change. I don't know when Climate Change was added to the name, but it is surely another sign of the times.

Selection of protected areas relies upon up-to-date inventories of the distribution of threatened and endangered plant and animal species in the province. These inventories are maintained and updated primarily at two major facilities, the Nova Scotia Museum in Halifax and the E.C. Smith herbarium at Acadia University in Wolfville. Both of these report to the Atlantic Canada Conservation Data Centre in Sackville, New Brunswick, which keeps records for all the Atlantic Provinces.

In 2023, Mary and I visited all three of these organizations.

We began with the Nova Scotia Museum of Natural History, on

Summer Street in downtown Halifax. This museum is dedicated to collecting and recording artifacts of cultural significance, as well as promoting Nova Scotia's natural landscape, and can trace its history back to the venerable Provincial Museum of 1868. It incorporates a vast herbarium of pressed and mounted plant specimens from Nova Scotia, dating back to those of Fernald and Macoun collected over one hundred years ago and even earlier. It was this herbarium that we were interested in, and it was the curator of botany, Sean Haughian, who was kind enough to show us around.

I knew that Sean was a highly-educated, widely-travelled, and accomplished field and research botanist, ecologist, and naturalist, and also an adjunct professor of botany at Saint Mary's University. The fact that he was the curator of botany at the N.S. Museum attested to his expertise in that field.

I was nervous going into the interview, afraid that Sean would be dismissive of an ordinary old man like me having the audacity to write a book about native wildflowers. I feared that I would be cowed and tongue-tied and forget to ask the right questions. At least I had Mary with me, and she's not afraid of anything.

A few moments with Sean allayed my fears. First of all, he was much younger than me, which gave me a sort of fatherly advantage. Second, he was talkative and answering questions before I had a chance to finish asking them. He was welcoming, friendly and eager to help me out. I forgot to be tongue-tied.

My first question for Sean was to get him to describe, in general, the role of the herbarium. The role, he said, was the same as that of the Nova Scotia Museum of Natural History itself, which is to preserve specimens that define the natural history of the province. In the case of the herbarium, it is botanical specimens that are preserved, while other departments of the Museum preserve specimens of fish, amphibians, mammals, insects, minerals, fossils, and artifacts of past cultures.

The plant specimens preserved in the herbarium consist almost entirely of collected plants, pressed and dried and mounted on sheets of quality herbarium paper. Each sheet is slipped inside a file folder for protection. These herbarium "'sheets" take up very little room, and there are thousands of them slid into big file cabinets in the climate-controlled herbarium rooms. Each sheet is identified as to species, date and location of collection, and name of collector. Each is numbered and recorded, and Sean knows how to find each and every one.

Plant collecting is alive and well in Nova Scotia in 2023. Sean supervises collectors in the field and collects in the field himself whenever he gets a chance. The purpose of collecting today is not so much the discovery of unknown species, but rather tracking the changing distribution of known ones.

Collected specimens are still preserved by pressing and drying between sheets of corrugated cardboard in a wooden plant press, just as it was done in the old days. A better way has never been found. In the herbarium workroom were presses full of recently-collected specimens drying.

Now, what I really wanted to find out was if the Museum of Natural History conducts any native plant classes or workshops for amateur botanists. Sean says that there are no regularly scheduled programs offered, but he and other museum staff are often asked to give talks or conduct outreach programs for garden clubs and societies, such as the Nova Scotia Wild Flora Society. The Wild Flora Society holds regular meetings at the Museum. Museum staff will also answer questions or identify mystery plants if a question or photograph is emailed to the museum website.

In the herbarium, Sean welcomes volunteers who are keen to learn the ins and outs of plant collecting, pressing and mounting dried specimens. Downstairs, the gift shop in the Museum sells a good selection of books about native plants and wildflowers.

Our next visit was to the E.C. Smith herbarium, on the campus of Acadia University in Wolfville. We had an interview scheduled with Alain Belliveau, the herbarium curator, for a Saturday afternoon, the day of the annual native plant sale, which we were to find out is a really big deal.

We planned our trip and booked our campsite weeks ahead. The plan was to drive to Wolfville from Antigonish on Saturday morning, early enough to be sure to make our appointment. Then we planned to camp three nights in Blomidon Provincial Park to do some hiking and wildflower exploring. We thought we would finally see the red trillium in bloom.

Then came the fires.

None of us will forget the catastrophic fires that erupted in Nova Scotia in June, 2023, scorching the forest, destroying homes, and driving thousands of residents from their communities. On the weekend in question, we learned that the Blomidon park was still open for camping but that all campfires were banned. Significant rain was forecast, which might answer the prayers of firefighters and maybe campfires would be allowed in the park again. We decided to stick to the plan.

Since we were early, we decided to head to the Acadia University campus and attend the native plant sale. We were an hour early, but already there was no place to park. We ended up in a gravel lot far from where the sale would take place.

By ten o'clock, starting time, we had joined a long line of people filing into the lovely conservatory complex, looking to buy native plants. This building and conservatory are part of the K.C. Irving Environmental Science Centre, which also incorporates the Harriet Irving Botanical Gardens, about which we had heard great things.

The size of the crowd is what amazed us. People just kept coming and coming. No wonder there was no place to park. Most of the people seemed to be young, and many came with small children. Even so, older people and senior citizens were well represented. Everybody carried trays which they were filling with native plants in pots, some sold by University staff and some by private growers. Everyone was smiling and excited to be taking home little native plants for their garden.

Mary and I were stunned at how interest in native plants and native plant gardening has skyrocketed since we were growing and selling them in the 1990s. We grew good plants and sold them at the local garden centre and through the mail. People visiting the garden centre were curious about our plants, but not enough to spend any money. Alongside the spectacular commercial varieties of peonies, German iris, lilies and the like, I'm afraid that our wildflowers looked pretty insignificant. We printed a catalogue and sent a lot of orders out by mail, but we were not charging enough for the plants or for shipping. I guess the time was not right, and after a few years we gave it up.

Judging by the crowds at Acadia this morning, maybe we should try again.

We were supposed to meet Alain when the plant sale was over at 12:00. It had begun to rain a little and we thought we would spend some time exploring the highly-esteemed Harriet Irving Botanical Gardens.

The woods everywhere in Nova Scotia were still closed because of fire danger, and we were surprised to find that so were the Botanical Gardens. The courtyard of the building was open, though, and planted with mostly native plants. We spent some time looking around there, then went inside where we checked out the conservatory rooms.

We were still too early, so I decided to go into the lounge area to sit and look over the questions I was planning to ask Alain. Many conservation groups had tables set up and were dispensing information. At other tables people were selling books and T-shirts and things like that.

At one of the tables, I recognized Alain Belliveau talking to visitors. I had seen pictures of him on a CBC news story about surveying for endangered plants at the West Mabou Beach Provincial Park, the contested site of a proposed golf course. He was hard to miss, anyway—a black-haired and bearded young man surrounded by adoring plant lovers. At odd moments when no one was asking him questions, he clowned around with his pretty little dark-haired daughter.

As soon as I had a chance, I introduced myself and reminded him of our interview appointment.

The plant sale was to end at noon, but Alain had been persuaded to lead a tour afterwards around the courtyard and through the facilities of the Environmental Science Centre. Mary and I decided to tag along.

We learned on the tour that one of the goals of the Herbarium and Science Centre staff is to develop techniques to propagate difficult and endangered native plants in order to re-introduce them into the wild. The plant they were focused on this summer was the almost extinct Ram's Head Lady's Slipper orchid.

In the production area of the courtyard, they had built a shelter to mimic the natural conditions where the rare plant is found. In the conservatory growth chambers, Alain and the staff were growing seedlings that would eventually be moved into the shelter.

Native orchids are notoriously difficult to propagate from seed. They also are almost impossible to transplant from the wild, which would be unthinkable anyway. Alain and the staff were testing procedures and were having some success with seed. They were also growing plants from tissue culture, which could be another way to get them started.

Figure 153: Ram's-head Lady Slipper orchid

By now, it was raining outside, but inside we enjoyed an interesting tour. When the tour was over, the group disbanded and Alain took Mary and me into a quiet room of the herbarium for our interview.

I was a little unsure how to start. In the long and comprehensive tour we had just completed, Alain had answered almost all the questions I had planned to ask. We had learned that every summer the herbarium sends out collectors to monitor native plant communities and send samples back. As at the herbarium of the Nova Scotia Museum, collected specimens are pressed, dried and mounted on sheets of stiff herbarium paper. Just as in Halifax, I spotted a plant press full of specimens drying.

The herbarium at Acadia contains over 200,000 pressed and dried specimens going back to around 1868. The present-day E.C. Smith Herbarium, of which Alain is the curator, opened in 2002. It incorporates hundreds of cabinets full of specimens in climate-controlled rooms, as well as an extensive seed bank.

Alain's job is a varied one. In addition to supervising the comings and goings of thousands of sheets in the herbarium, he is involved in propagating living plants for the Botanical Gardens and for reintroducing plants into the wild.

Most of all, he likes getting out into the field himself. His dream is to contribute to protection of priceless natural heritage and biodiversity by discovering, surveying, and mapping threatened species and habitats, thereby forestalling thoughtless destruction by private or commercial development.

I could think of only three questions appropriate to this book that Alain had not already answered.

The first: Do the Herbarium and the Botanical Garden conduct workshops or courses to teach aspiring amateur botanists to identify and appreciate native plants? Not exactly, answered Alain, but herbarium staff offer regular tours of the Botanical Garden in which many native plants are featured. Alain leads native plant outings and conducts workshops with the nearby Blomidon Naturalist Society.

The second: Will Herbarium staff help an amateur botanist identify a plant species they have discovered? Yes, says Alain, willingly, if someone sends or brings in a photograph or even a fresh specimen. The photograph is best. Fresh specimens must be cut off —never pulled—and must not be taken at all if the plant seems scarce. Also, fresh specimens must be kept cool in plastic bags to prevent wilting and melting into an unrecognizable mass.

For most discoveries, this discussion is probably moot anyway, unless you just want to meet Alain, because today there are good field guides and easy ways to identify plants through the internet. I will talk about these later.

The third: Does Alain need and welcome volunteers to help out in the herbarium? Definitely, he answers. They can help with pressing and mounting specimens and learn about our native plants at the same time. There is also a need for volunteers to help care for the plantings and the grounds of the extensive Harriet Irving Botanical Gardens. This could be an opportunity to gain some gardening expertise while becoming familiar with native plants growing in a natural setting. The Botanical Gardens website explains how to apply for volunteer work in the gardens or herbarium.

By the time we finished our interview with Alain, it was raining hard off and on. In view of the fires burning out of control in the province, no one wanted this rain to stop.

We drove to the Blomidon Provincial Park and managed to get our tent set up during a lull in the rain. Though it was raining, there still was no campfire allowed and not even any hiking. We ate a cold supper in the car and went to bed early.

During the night the rain really picked up and water found its way into the tent. Sunday, we crawled out of the tent and got dressed and freshened up in the comfort centre. We drove into Canning and had breakfast, then back into Wolfville. We spent the day half-heartedly touring garden centres and ate supper at the A&W.

Back at Blomidon, it was raining harder than ever. We spent another night in damp sleeping bags and bailed out the next day, cancelling our last night. A wet tent, no campfire, and no hiking. Amateur botanists can only take so much.

Someday we will go back for the hiking and also to visit the Harriet Irving Botanical Gardens, which we missed and of which we have heard so many good things.

Our next destination, a few weeks later, was to visit the Atlantic Canada Conservation Data Centre (ACCDC) in Sackville, New Brunswick, and to meet Sean Blaney the director. We had learned from Sean Haughian and Alain Belliveau that the results of all botanical surveys and collecting done across Nova Scotia each summer end up at the Conservation Data Centre.

As the name implies, the function of the Data Centre is to record and map locations of sensitive habitats and plant communities. The AC in ACCDC means that this Centre in Sackville performs the same services for all the provinces in Atlantic Canada, including Newfoundland and Labrador. We would soon learn that there was much more.

This time, since Sackville is so far from home, we had booked a campsite at Amherst Shore Provincial Park, just across the border in Nova Scotia. Our plan was to drive early to Sackville from home in our good clothes, so we would create a good impression and not look as if we had just crawled out of a tent. After the interview with Sean Blaney, we could go camping and get back into our out-door clothes.

Again, we arrived in Sackville early, which was a good thing. It took us a few trips around town to find the Mount Saint Vincent University campus, then a few trips around campus to find the AC-CDC building.

The building was a venerable old wooden house on the grounds of the mostly stone and brick Mount St. Vincent U. There was no sign saying Atlantic Conservation Data Centre, so I went inside to make sure we were at the right place.

Up a steep flight of stairs inside I found Sean's office, identified by conservation posters on the walls. Sean wasn't in so we went into town for lunch.

After lunch, still a bit early for our appointment, we returned to the ACCDC building and amused ourselves sitting on a stone wall in the shade and watching people go by and wondering if one of them was Sean. I had booked the interview three or four weeks back, had heard nothing since, and was a little worried that Sean would forget.

I shouldn't have worried. We found out that Sean doesn't forget anything.

Precisely at 1:00, the time at which we were to meet, one of the two men we had been watching peeled off and shot up the steps of the wooden building two at a time. That had to be Sean.

This was our cue to go in. Sean was at the top of the steep stairs inside to greet us. In his office, the most prominent things were a big computer screen and, yes, a stack of wooden presses full of plants drying.

We began by talking about the old house, which was being thoroughly renovated downstairs. Sean told us that the University was hinting that they wanted it back and the ACCDC would probably have to move.

While Mary and Sean talked about the house, I checked him out. He was another of those young, tanned, and athletic-looking botanists like Sean Haughian and Alain Belliveau. The three of them look as if they might belong to the same mountaineering club.

When I began listening again, we were talking about plants.

Sean was explaining what they do at the Data Centre. What they do is collect data from across Atlantic Canada. Every province in Canada has its own Data Centre, but the Atlantic Provinces are small and share the one in Sackville.

In the summertime, Sean and his colleagues work in the field, surveying and collecting and sending information back to the Data Centre. Many other collectors working across Atlantic Canada do the same thing. The Centre employs GIS technicians, whose job is to record and map all this data. Eventually these maps can be consulted to see if a proposed private or commercial development is likely to threaten sensitive habitats or plant communities.

The maps can likewise be used to select particularly sensitive areas needing protection in the form of nature reserves or protected wilderness areas. The first time I encountered the name Sean Blaney was as a co-author of a study of the unique flora and fauna of threatened gypsum areas in Nova Scotia. One of these was the riverbank near us that was, as a result of this study, set aside as a nature reserve. Proof of the importance and value of these sorts of studies, and of the Data Centre itself.

Now things get complicated. The Data Centre, in addition to tracking almost unbelievable amounts of botanical data, also keeps track of birds, animals including reptiles and amphibians, butterflies, and other insects of concern throughout Atlantic Canada. This is possible because, in this day of computerization, all this data doesn't require a building to store it in. Sitting in the office with the desktop and the plant presses, Sean has it all at his fingertips.

Computerization, specifically the smart phone, has also revolutionized collection of data useful to the ACCDC. It is no longer only professional botanists who send in data. You can do it, too. In fact, observations and photos taken by amateur botanists and naturalists, or just people enjoying the outdoors, are today an important component of conservation data.

Sean explained to us the principles and significance of the new (to us) iNaturalist phenomenon, which had its beginnings at the University of California in 2008. iNaturalist is now used by people all over the world and plays a double role.

First of all, a photo of an unknown plant or animal submitted to iNaturalist is immediately identified for you. This, obviously, is a great help when you are out in the field. Sean says that this initial identification is about 80% accurate, and depends to a large extent on the quality of your photograph.

The photograph is then examined by experts to confirm the identification. Sean is one of these experts, and says he has been called upon maybe a thousand times to verify identities.

Once the identity is confirmed and deemed a noteworthy find, it is sent to the iNaturalist database, from where it is available to conservation planners and researchers anywhere in the world. Observations pertaining to Nova Scotia or the other Atlantic Provinces come to the ACCDC in Sackville, and are recorded and mapped.

A staggering amount of data is obtained this way to augment data gathered by professionals. The provinces of Nova Scotia, New Brunswick, and Prince Edward Island have set up provincial iNaturalist Rare Species sites where those wishing to help track designated endangered species can report sightings.

Mapping cell phone observations is made possible by the fact that GPS coordinates are embedded in the photo, making it possible for technicians to pinpoint the exact spot where the picture was taken. This is a great boon for the conservationists at the ACCDC, and helps considerably to grow their database.

The distressing part to me is that not only scientists and professional conservationists can determine where you took the picture —anyone can. I am all for helping conservation, but I am protective and jealous of my favourite places. I don't want most people to know where they are. That is why they are my favourite places— they are unspoiled and known to very few. Sean says that unscrupulous native plant collectors have been known to swoop in and dig up plants at sites they learned about on iNaturalist. God forbid!

There is some protection for those using iNaturalist. By clicking the right button you can either block your GPS coordinates completely, or obscure them so only qualified scientists or planners can use them. I haven't begun to participate in iNaturalist yet, but this would give me some feeling of security.

We concluded our interview with Sean Blaney after an hour or so, deeply impressed and thankful for what he, Sean Haughian, Alain Belliveau and a host of others we haven't met are doing to protect biodiversity in this part of the world.

From Sackville we drove to Amherst Shores where we camped. It is a very nice park and I talk more about what we did there in the chapter on parks and trails.

I started this chapter thinking I would list all the aids I could think of to help you learn plants and identify your finds. Those include internet apps like the aforementioned iNaturalist. There are similar ones on the internet, though I think that iNaturalist is the only one that enters photographic observations into a biodiversity data base.

Also included is an old-fashioned thing called books. Mary and I use both the *Peterson Guide to the Wildflowers of Northeastern North America* and *Newcomb's Wildflower Guide*. Both are good. The best thing about the field guide is that it is slow, and you learn many other plants while you are flipping through looking for the one you want to identify.

We also use *The Flora of Nova Scotia*, which, if the plant has conspicuous flowers and you are familiar with botanical terms, will lead you to the right answer. The Flora uses the dichotomous key system, whereby you pick one of two choices, which leads you to two more choices and so on and so on until the plant is identified.

Sean Blaney showed us an interesting site on the computer called Go Botany, which uses a pictorial version of the dichotomous key to help amateurs identify plants.

Native plant and naturalists societies offer lectures, workshops, tutorials, botanical field trips, and companionship with other plant lovers. Some of those we have heard of are:

- The Nova Scotia Wild Flora Society. This is a long-established society with a large membership, and hosts activities outdoors in the Summer and indoors in winter. It meets once a month at the Nova Scotia Museum of Natural History.
- The Blomidon Naturalists Society
- Halifax Field Naturalists
- Cape Breton Field Naturalists

And finally, regional garden clubs welcome members interested in any aspect of gardening, including native plants.

Figure 154: Field guides

14: To tell or not to tell

Figure 155: Canada lily

Let's suppose you are out exploring for plants and you find something very special, maybe a rare plant, or a particularly rich community of them. These things do happen.

Now you are faced with a dilemma. Do you keep it a secret, or do you tell? And if you tell someone, who will it be?

Two personal examples:

The first: When I worked at the Pleasant Valley Nurseries in Antigonish, we spent much of the summer travelling around the region doing landscaping. For a couple of days we worked planting trees and shrubs around the house and barns at a local dairy farm. The farmer and his wife were very nice people and loved plants. We often saw them at the garden centre.

The driveway to their farm turned off from a gravel road that followed the river. The driveway was on the west side of the road. Directly opposite was a pocket-sized hayfield squeezed up against dense alders, chokecherries and wild roses along the river. The farmer had been working to enlarge the hayfield, and had mowed a couple of swaths through the brush.

Growing from the stubble, and clearly visible from the truck as we made our turn, were four or five tall, stately Canada lilies in full bloom. The Canada lily is in a class of its own when it comes to native wildflowers. It stands three or four feet high with dozens of large, flared, bell-shaped, orange or yellow flowers on each plant. It is partial to the river floodplain, and at the time of the first colonists, before the floodplains were cleared, the show must have been astounding. The shopping mall in Antigonish is built on what was formerly the broad floodplain of the West River, and I'm told that at one time Canada lily was plentiful there.

Because it is so big and beautiful, the flowers were heavily picked in the past, and now the floodplains have been cleared and plowed. The Canada lily has nearly vanished from Antigonish County. Four or five standing together, like those we had discovered, was a rare and unforgettable sight.

Because of where they were growing, in the newly-mown stubble at the edge of the hayfield, it seemed the lilies were in imminent danger of being cut down. Since the people we were working for were fond of wildflowers and nature, I was sure they would protect the plants if I tipped them off to their presence.

Before we had finished our work, I had a chat with the farmer and his wife about the Canada lily and how rare it was. I cautioned them to take care not to mow them down. They were excited to learn they had Canada lilies on their own land and promised to look after them. Heading back to the garden centre, I felt very good about myself, having discovered the lilies and made sure they were protected.

Several years later, I began to wonder how the Canada lilies were doing and went back to find out. I guessed that there would be lots of them by now. It was August and they should be in bloom, which would be quite a sight.

When I arrived at the spot by the river, where I expected to see the lilies at the edge of the hayfield, there wasn't a trace. Where the lilies had been, the bushes had all grown back. I walked up and down the length of the field, and peered into the bushes. No lilies. What had happened?

It took some thinking but I figured it out. I felt pretty stupid. I had been afraid the farmer would mow the lilies down and kill them, but that wasn't it at all. Thinking back, those lilies were very large. The year I discovered them couldn't have been the first year they appeared, they were too big. They must have been growing happily in the stubble for several years. Evidently, the farmer had been mowing the brush once a year, probably at a time when the lilies were dormant. When the weather warmed in summer, the lilies shot up and bloomed with no competition from the bushes. By the time the brush was mowed again, the lilies were done for the year.

It could have gone on like that indefinitely, but when I alerted the farmer to the lilies, he quit mowing altogether. The brush grew back in, and the lilies disappeared. A classic case of good intentions gone wrong. It would have been better to keep my mouth shut.

Example two: Near the farm we once owned, there was a woods road, a logging road that is, that was built by a logging company through mostly-mature spruce forest to get the wood out. Over the next year, tree harvesters worked night and day, and the spruce was all taken down and trucked to the sawmill.

Virtually everything on the east side of the road was clearcut and gone. The land on the west side was apparently owned by someone else and was untouched. This piece of land was mostly old hardwoods and more or less followed the river a half kilometre away. I didn't know who owned that land.

The soil beneath the old hardwoods could be considered river intervale, and was rich. The growth of bloodroot, dutchman's breeches, trilliums and the other intervale plants was extensive. In June, the fiddleheads came up everywhere, and we went there to pick.

Figure 156: Fiddleheads for pickin'

At the time, Mary and I were operating a wildflower nursery, and this piece of land was a treasure trove of rare species. We went there to learn our plants and collect seed. Besides the bloodroot, dutchman's breeches and trilliums, we took seed from toothwort, yellow violet, twinflower and the rarer Solomon's seal and dwarf ginseng. All these plants were indicative of old forest and rich soil.

Then, one winter, whoever owned the land began to take down the old hardwoods. In Spring, we were heartbroken to find most of the trees gone and the ground torn up and run over.

The cutting stopped, and we were happy to discover that a small, triangular piece of the forest was intact, in a corner where two woods roads came together. This piece was maybe only a quarter of an acre, but had a good population of rare plants.

The cutting didn't resume the next year, or the one after that, and I was convinced that the little piece of forest was just too small for the landowner, whoever he was, to bother cutting. I considered trying to find out who owned the land, pointing out that it was a special place, and asking if he would leave it alone. I was too busy to do this, though, and told myself that it was probably best if I kept quiet and didn't draw attention to the spot.

Well, you can probably see it coming. Mr. Whoever He Was hadn't forgotten about that spot and he came back and cut it.

The worst part was that I did eventually find out who owned the land. He was a neighbour not far away and a good, reasonable guy. If I had asked him to spare that little piece of land he probably would have done it. This time, I should have spoken up.

To tell, or not to tell, that is the question, and it depends on different things. For sure you will tell your partner, and any of your immediate family, children, or friends who might be interested, and take them to see it. Beyond that, my instinct has always been that the surest way to protect a discovery is to keep it secret. Maybe this is a bad case of wishful thinking.

The objective, considering the plight of threatened species in Nova Scotia, is to protect the plant and its environment. If you own the land, this is quite possible. You keep it to yourself and only show it to whomever you select. If you don't own the land, you get into the grey area. It depends on who does.

The places Mary and I visit, we feel are secure in a shaky sort of way. This is because they are places that are not likely to be "developed", or else places that are protected already. The exception is the little birch forest that I have described as an elfin forest, near our cabin in Cape Breton.

This forest is on a property next to us, and is a piece of what was once a large farm. The land was held by the descendants of those who farmed. They had a little shack in the trees by the shore, with a barbecue and a few chairs near the beach, but we seldom saw them.

That part of the old farm was mostly a big, grassy field. There was always the danger, to our way of thinking, that one of the owners might begin construction of a big house in this field and put an end to our privileged privacy. This was unlikely, though, because there were problems with access from the road, and getting in electrical power.

Years passed, nothing happened, and we quit worrying. Then For Sale signs went up and came back down again. The place had been sold. No one on the road was quite sure who owned it now. Rumour had it that it was a millionaire from Chicago.

First thing in the spring, a dozer and excavator appeared and began to put in a driveway. This driveway couldn't be built straight up through the field, which would have been the logical way to go, because the access was blocked by a piece of private property. That meant that the new driveway had to go up through the forest next to the field, through a cattail swamp and beaver ponds, and then up close alongside our elfin forest.

Once he got into the swamp and beaver ponds, it was slow going for the excavator. He had to make crude bridges here and there as he went along. I couldn't understand how he could even work without getting stuck.

For a long time, the excavator was stalled at a particularly deep channel of water. We hoped that maybe the millionaire had given up and abandoned the idea. Failing that, given the environmental damage taking place, we hoped the Department of Environment would shut him down. We considered reporting him ourselves, but we couldn't be that unneighbourly to someone we hadn't met.

Later in the summer, work resumed and the drive was pushed through to the top of the bluff overlooking the ocean. So this is where the mansion will be built, we figured, except that the new drive was useless. In the driest days of summer, it was still too muddy and wet to get through, even in rubber boots. Neighbours tried it on ATVs and couldn't make it.

The machines went home, and no one ever showed up to use the driveway. It has been a couple of years now and the driveway scar across the swamp has filled in again. Our elfin forest was narrowly bypassed.

We have never seen nor met the owner of the land. If we do, and if he resumes work on his property, we will certainly talk to him about saving the little birch forest. Chances are he bought the land to get back to nature and would be all for it. Even so, if the driveway was finished, the fox would leave his den, and things would never be the same.

The other places we visit to see wildflowers are a little less precarious. The bogs and the gypsum cliffs are seemingly safe because there is no way anyone can build or make money off them. Of course, some corporation could decide to take the gypsum for wallboard or the bog moss for peat, but we trust it won't happen soon.

Both areas are sensitive, and we only take certain people with us to see the flowers. I'm quite sure that I know who owns the gypsum, but I wont tell him there are lady's slippers up there. The cliffs are crumbly and I think it would only take a handful of people wandering around aimlessly to ruin the place.

Our spring ephemeral and fiddlehead fern sanctuary along the river seems secure. It hasn't been tampered with in the thirty years we have been going there. It is situated between a long-established hayfield (now cornfield), and the river. I don't believe the farmers have any interest in expanding their field, and are, in fact, probably prevented from doing so by stream bank protection regulations. Some local people visit this riverbank to collect fiddleheads or to fish, but usually we have it to ourselves.

It seems selfish to keep quiet about a good thing. Maybe it is an old-man thing. All my buddies agree that you don't want too many people finding out about your favourite places. I think you might call it the fishing hole complex.

The fishing hole complex extends to all kinds of things. As the name suggests, a fisherman shudders at the thought of anyone finding his special hole. People who harvest fiddleheads or cranberries or chanterelles keep their best spots secret. If you know how to get to a little known waterfall you keep that secret, too.

I bet that people in the cities keep secrets, like the exact location of their favourite little restaurant, or the secret knoll from where they can watch the rock concert on the commons for free. In the context of this book, it is places that are rich in wildflowers that we wish to keep secret. Maybe it isn't just an old-man thing. Maybe it is human nature.

The trouble is that, in 2023, I am not sure if there is still such a thing as human nature. Everything is computerized. People's lives and thoughts are on display on social media and the internet. Experts tell us what to do and what to think, and soon AI is going to do the thinking for us. Today the urgent thing is to open up and save the planet and the wild things we share it with. Maybe it is time to tell.

Sean Haughian, Alain Belliveau, and Sean Blaney are unanimous: they want to know what we have found. For the maps they compile of endangered plant populations and habitats, they depend heavily on discoveries reported by amateur naturalists. This no longer requires pressing and mounting dried specimens. It only takes a photo on the cell phone.

Photos can be taken and sent directly to the Nova Scotia Museum, or the E.C. Smith Herbarium, but they all end up at the Atlantic Canada Conservation Centre with Sean Blaney. The ACCDC also welcomes photos sent directly to them if you follow the procedure on their website.

To simplify things, though, Alain and the two Seans recommend that you take advantage of the iNaturalist app. Data obtained this way may be instrumental in the selection of natural areas to be protected, so we should take it seriously. If you are concerned, like I am, that too many people will invade your secret spot, click on the button that partially masks your GPS location and then, I guess, trust your fellow man.

So now back to the dilemma of "to tell or not to tell". In 2023, with climate changing and all God's creatures under siege, maybe it is time to listen to the botanists, drop our guard, and tell. Perhaps information obtained from our observations will result in even more reserves and refuges for wild things in years to come.

15: Wildflower gardening

For anyone who loves native plants and wildflowers, and gardening, it is natural to want to get wildflowers growing at home. It is not a simple matter, though, of digging up plants from the wild and taking them home. As a matter of fact, this should almost never be done. Many native wildflowers do not transplant readily, while others are endangered and should never be dug.

It is necessary to become familiar with the plants, and the growing conditions they require, before attempting a wildflower garden. This is not to say that wildflower gardening is difficult, but as with a conventional perennial flower garden, each species has specific requirements and challenges.

Mary and I worked and met at a retail garden centre in Antigonish. Work was insanely busy at the garden centre during the months of May and June, just when the native wildflowers were blooming their best. Between long shifts at the garden centre, and getting a vegetable garden started at home, and looking after animals and kids, we barely noticed the fleeting bloom of wildflowers in the woods all around us. We discovered it by accident.

I had, after driving past it for a number of years, I am ashamed to say, discovered that there was an extensive growth of bloodroot at the foot of the brushy roadside bank where it joined the hayfield below. It was midsummer when I noticed it. It had long since finished blooming, but I recognized the leaves from pictures in a book. I slid down the bank to break off a leaf and, sure enough, blood red sap.

I made a point of looking over the bank the following spring when the bloodroot was in bloom. It was so spectacular that I couldn't understand how I had missed it other years.

And not only was there bloodroot at the foot of the bank, but dutchman's breeches, yellow violet, and herb robert, a tiny pink blooming geranium. In the fall, the farmers decided to widen their field, and plowed it all up.

It just so happened that I had built a tall stone planter at home to hold a flowering dogwood, and I needed soil to fill it. When the farmers widened their field, they left some good-sized piles of rich-looking soil. I helped myself to some of this—I knew they wouldn't mind—filled the planter, and put in the dogwood.

When spring came, the dogwood came to life as I was expecting, and so did bloodroot, Dutchman's breeches, yellow violet and herb robert that I wasn't expecting at all. They must have been in the soil I took from the pile in the field.

Figure 157: Wild yellow violet in the garden

This is how Mary and I got started in wildflower gardening.

After we had caught the bug, we began to study and search out wildflowers at any opportunity. We joined the Canadian and the New England wildflower societies and the Nova Scotia Wild Flora Society, and received their magazines and newsletters. Books on wildflower gardening came mostly from the U.S.A., and dealt with our common native wildflowers as well as many others not quite native in Nova Scotia.

Some of these, such as cardinal flower, bee-balm and cone-flower, were easy to obtain and tempting to grow. It was clear that flowering plants don't recognize the Canada/US boundary line so neither did we. Anything found in Northeastern North America was fair game.

It was very unusual in those days to find native plants for sale at garden centres, and that included our own. Occasionally we had pots of half-dead trilliums or hepatica shipped in from Ontario, but they were often all dead soon after planting. We decided to grow our own.

Both the Canadian and the New England wildflower societies gathered and sold wildflower seed. We ordered a wide variety, and seeded them into flats.

Growing wildflowers from seed is not as simple as annuals that begin to grow as soon as they are sown and watered. Many wildflower seeds require pre-treatment, most often several months in damp soil at near-freezing temperatures. This is called "stratification", and can be accomplished by mixing seeds with slightly damp peat moss in a plastic bag and keeping them in the refrigerator for a few months.

In nature, stratification takes place naturally when seeds fall to the ground in summer and fall, then spend the winter in damp soil, often under snow. The genius of this system is clear when you think about it. If wildflower seeds germinated as soon as they fell to the ground, the young seedlings would never make it through winter. Stratification delays germination until the ground warms up in spring, when seedlings have all summer to grow.

Many of the wildflower seeds we purchased germinated, and our inventory began to grow. We planted as many as we could in new wildflower beds we were creating and, reluctant to throw them out, potted up the extras.

Soon we had quite a collection of potted wildflowers and gave many away, then began to wonder if maybe we could sell them. We were quite aware of the difficulty of finding good wildflower plants at garden centres, so we began to label ours and take them in for sale. Eventually we had a respectable assortment of wild-flowers for sale.

They weren't selling hand over fist. In those days, not many gardeners came looking specifically for wildflowers, pots of which mostly lacked the visual clout of common perennials. Still, we were having fun, moving along our plants, and breaking even. We decided to start a business.

We named our business Borealis Wildflowers, and got it registered. Our plan was to continue to sell plants at the garden centre, but also print a catalogue and sell through the mail, which we did. Meanwhile, we continued to tend our gardens at home and propagate new species of wildflowers.

We operated the business for several years until we had to accept that we weren't making any money the way we were doing it. Considering the time we put into it, we weren't charging enough for our plants, or for shipping. We were still just breaking even. Along with this, our children were getting older, we were short of time, and Mary was nursing me through cancer treatments.

We had learned a lot running Borealis Wildflowers, met many people and made friends, but sadly, we let it go. Here, though, are some interesting things we learned.

Wildflowers, once planted, tend to find their own niche and hang on to it. In fact, the most amusing thing we discovered about wildflowers is that if they don't like the spot where you plant them, they will move. More than once, a plant struggled unhappily and died in the spot we had chosen, only to reappear, presumably from seed, and thrive in a completely different place—even years later.

We had a small farm with a variety of habitats for a wildflower to choose from, ranging from tended bark-mulched shrub beds to rough and snaggly woods. An unexpected colony of wildflowers could appear anywhere, started, I guess, from seeds scattered by unhappy plants from a first clumsy planting. These colonies, which had chosen their own spots to grow, invariably thrived and spread.

As the years went by, wildflowers actually went wild, intermingling on their own and requiring very little interference from us. That is the essence of wildflower gardening—getting plants started and then letting them go. From time to time, a new wildflower appeared that we didn't remember planting and that must have somehow arrived on its own.

To be honest, not all of our wildflower plantings were trouble-free. We sold or planted three categories of wildflowers: woodland, wetland and meadow plants. The wetland and meadow plants were largely fast-growing, robust and tough. They could fend for themselves and almost always succeeded in plantings.

Woodland plants, on the other hand, were not so vigorous. In the wild they grow under the leafy trees of the forest, where the soil is loose and rich and the plant community is settled and stable. This was not exactly the case where we planted on our farm. We could duplicate the soil and shade conditions, but the plant community was not at all stable.

Wildflowers had to cope with the tough garden weeds that come with civilization. Mary and I spent plenty of time on our hands and knees, pulling weeds that threatened our delicate woodland pets. One planting in particular featured a hollow log set among trout lily, Dutchman's breeches, hepatica, maidenhair fern and similar treasures. These were happy in this spot, under old apple trees, and were slowly spreading, but needed constant protection.

I had carelessly let Dame's rocket get started and it wanted to grow four feet high and take over. Then there was an annual crop of lamb's quarters and hemp nettle that also wanted to take over. They ran right over our little six-inch woodlanders. It took a lot of weeding each summer to keep them safe. In areas a little deeper into the woods, though, the weeds weren't bad and wildflowers established themselves without asking our permission.

Our most successful wildflower garden was on the north side of a tall hazel hedge that screens the house from the road. This was a situation most woodland plants were looking for. Before the hedge leafed out in spring, there was plenty of sun coming through the bare branches; then, after the hedge leafed out, plenty of shade.

Bloodroot had already begun to pop up in various places on our property, and we moved some into this garden. We placed large, attractive rocks in strategic places and began to plant trilliums, Virginia bluebells, Solomon's Seal, and other promising plants we wanted to try out. At the time, we were propagating native ferns from spores and filled in between the wildflowers with young ostrich ferns. We had plans of eventually harvesting our own fiddleheads right at home.

Planting the ferns was a lucky decision, and the reason this garden was so successful. The ostrich fern spread quickly in the rich soil behind the hedge. It grew in scattered clumps, so did not crowd out the wildflowers. In fact, the spring wildflowers finished blooming before the ferns came to life. When they did come to life, they shot up three or four feet high and completely shaded the ground, making it impossible for weeds to grow.

Fearing to upset this precious equilibrium, we left those ferns alone and continued to get our fiddleheads at the river.

At one end of this wildflower garden, where the ostrich ferns were absent, we had good luck with other wildflowers and ferns, including Starry False Solomon's Seal, wild geranium, and maidenhair fern, which is almost extinct in Nova Scotia but grew vigorously in this spot.

When we had a flat of wildflower seed that hadn't germinated after a couple of seasons, we gave up and emptied it into the wildflower garden. Sometimes we were surprised to find something growing that must have been dormant in one of these flats. An example was red trillium, which I definitely hadn't planted, and which was spreading nicely.

There is no thrill quite like discovering a community of wildflowers in its home in the wild. Unhappily, the wild is threatened on every side. Farming and forestry and housing developments remove forest and plow up land with no let-up. Undisturbed areas of woodland are smaller and smaller every year and harder to find. Wildflower gardening is a way to observe and appreciate threatened species that were common at one time. You may find, as we did, that some of the species you plant spread around your property and up and down the road, and aren't as threatened anymore.

Digging wildflowers from the wild is unacceptable, unless the

Figure 158: Black-eyed Susan with swamp milkweed

plants are facing destruction from some development. If plants are obviously widespread and common, it is probably okay to take a couple for your own garden, but not for sale. At Borealis Wildflowers we started our plants from seed we had gathered or ordered from wildflower societies. If we had enough plants in pots, we divided them up and started new plants that way.

If you want to plant wildflowers at home, read up on the plant and put it in where you guess it will be happy. Don't be surprised if it moves on you. Woodland wildflowers grow up and bloom before the leaves come out on the trees. They combine well with ferns, which cover the ground after they are finished.

I liked my planting centred around the hollow log. Wetland and meadow wildflowers almost always like full sun, or close to it, and combine well with clump-forming grasses.

Loosen up the soil, and mix in naturally-occurring things like peat moss and rotted leaves, or weed-free compost. Bone meal is good fertilizer for wildflowers. Avoid raw barnyard manures that are probably full of weed seed.

Once you are hooked, wildflower gardening is habit-forming and addictive, so beware. In contrast to the understated, God-given beauty of wildflowers, man-made hybrids begin to look pretty awkward—Somewhat like the difference between a doe and a cow.

Conclusion

Mary and I have now spent three memorable summers out-doors, gathering material for this book. We hiked and camped and were on the lookout for beautiful wild places and native wild-flowers. When driving, we did our best to stay on the road as we craned our necks to see what was growing in the ditches.

Maybe we neglected our lawn and our gardens and our friends and our kids, but the hours we spent along the rivers, in the bogs and barrens, in the gypsum or the forest, or by the falls were good for our souls. We were looking.

Sometimes we were looking for a certain species to photograph for the book. Sometimes we were looking for a promising site. Always we were looking for anything new, and always we found something. Always I was looking for orchids or the fragrant fern. Always we were entranced by the simple beauty of the natural world.

We proved over and over again that there is joy in looking. That is the message we hope to have conveyed in this book.

Nova Scotia is an incredibly diverse province and, by world standards, incredibly wild. As I write this, there are three bald eagles soaring in the sky overhead. I just returned from an aban-doned pasture next door, where I discovered three more species of wild orchid as I crawled around on my hands and knees picking blueberries. When I get tired of writing I will go down to the shore for a swim.

Believe me, very few people in the world have what we have. Here in Nova Scotia, these things are in reach of everyone. Even those living in the heart of Halifax, our only large city, can be out in unspoiled nature in a very short time. Learning to recognize and appreciate wild plants, and learning their names, are enjoyable reasons to get out. Another objective of this book.

We all know the health benefits, both mental and physical, of fresh air and exercise. Plant hunting offers plenty of both. They tell me that it is impossible, but I swear hiking trails have more uphill than downhill. Hikers have strong legs. No place in Nova Scotia is too far above sea level, so there is a lot of oxygen in the air, which is a bonus. More reasons to get outdoors.

While we do have considerable wild land in this province, it is fragmented, and constantly threatened by more or less dubious private and commercial developments. I am thankful that the Nova Scotia government has committed to protecting 20% of our wild-lands by 2030. Significantly large tracts of land have already been set aside to protect them from development.

The federal government is doing its part as well. For example, it has very recently recognized three remote islands in Nova Scotia, namely St. Paul, Country, and Isle Haute, as valuable bird sanctuar-ies and will protect them as National Wildlife Areas.

In Nova Scotia, the Department of Environment, specifically the Protected Areas Branch, works with the Nature Conservancy of Canada and non-governmental organizations such as the N.S. Nature Trust and the Atlantic Canada Conservation Data Centre to identify regions of unusually valuable biodiversity needing protec-tion. Steps are then taken to set these areas aside, or purchase them if they are privately owned. Once secured, they may be de-clared protected areas.

Many private citizens owning or living on properties with out-standing natural beauty or biodiversity, are choosing to sell or donate their lands to be protected for their children and future generations. The Nova Scotia government, in tandem with the Nature Conservancy of Canada and the Nova Scotia Nature Trust, offers generous options whereby your property may be acquired by the province or designated a nature reserve or wilderness area if you maintain ownership. This protection is binding on all future governments or purchasers of the property.

Anyone interested can start by contacting the Protected Areas branch of the Nova Scotia Department of the Environment.

Summer 2023 is winding down. Mary and I have investigated every interesting environment we had hoped to, and taken lots of photographs, the best of which are in this book. I haven't yet found the fragrant fern, but I tried. I did find six native orchids that I could identify, and a couple that I couldn't. That only leaves about thirty species left to go.

The orchids and the fern are only a side show, of course, an excuse to wander among the splendours of nature. We hope you may do some wandering of your own and find as much joy as we have.

Figure 159: Goldthread in bloom

Bruce Partridge

Images

Bruce Partridge

Acknowledgements

I would like to thank my friends Henri Steeghs and Jeffrey Parker, who helped me with encouragement and photographic expertise.

Also, Sean Haughian, Alain Belliveau, and Sean Blaney, botanists extraordinaire, for their information and advice; and Andrew Wetmore, who edits and polishes up my writing to make me look good.

The Carl Linnaeus portrait on page 20 is from an online article, "His Career and Legacy", from the Linnaean Society.

The image of Captain Cook's ship *Resolution* on page 21 is derived from a drawing held by the Royal Geographic Society.

The drawing of Titus Smith on page 22 is from an online article discussing *Exploring Paradise: Titus Smith in Nova Scotia 1801-02*, edited by Warren C. Reed and self-published in 2017.

The photo on page 37 of Merritt Fernald and Bayard Long in Virginia in 1942 is from an online article about Fernald as 'the father of Newfoundland botany'.

The image of Margaret Sibella Brown on page 43 is from the Beaton Institute Archives.

The image on page 44 of (l to r) Charles Weatherby and colleagues Ludlow Griscom and Alfred S. Goodale is from the archive of the Gray Herbarium at Harvard University.

Bruce Partridge

About the author

Bruce Partridge and his wife, Mary, live in St. Andrews, Antigonish County, Nova Scotia. Both are graduates of the Ornamental Horticulture program at the old Nova Scotia Agricultural College in Truro. They met at the Pleasant Valley Nurseries in Antigonish, where they worked together for decades. Together they have long been interested in native plants.

They enjoy Summer jaunts into the wilds looking for new ones. In the 1990s they operated a nursery of their own, propagating and selling wildflowers at the garden centre and through the mail, and are tempted to start up again.

His first book, *Borderline Hardy in 5b*, is also available from Moose House.

www.ingramcontent.com/pod-product-compliance
Lightning Source LLC
Chambersburg PA
CBHW061141120626
46546CB00005B/1886